MW01488835

Thank You for...

For everything —

I appreciate your

hospitality, friendship and Merci —

— Enjoy —

Sincerely Russell

AKA
Proud Father

This volume is a signed, first edition of

WILD MAN
A MOSTLY TRUE MEMOIR OF A MISSOURI CATTLEMAN

Each copy is signed and numbered by the author and the Wild Man.

This is number **346** *of* **385** .

WILD MAN

WILD MAN

A MOSTLY TRUE MEMOIR OF A MISSOURI CATTLEMAN

BY SIERRA SHEA

FARM HOUSE PRESS

Printed in the United States of America.
Book design by Tim Murray, paperbackdesign.com/books
Edited by Cynthia Epp and Allison Hugill

Dedicated to my grandfather—
you are a wild man!

In memory of
L.W. Angell and
Louise Wisman Angell.

TABLE OF CONTENTS

"In telling the story of my father's life, it is impossible to separate fact from the fiction, the man from the myth. The best I can do is to tell it the way he told me. It doesn't always make sense and most of it never happened… but that's what kind of story this is."

– Will Bloom, *Big Fish*

PROLOGUE

LUTHER'S TWO LIFE GOALS

"Have you kids heard the story about Papa's life goals?" my father asks.

With my cousins and my sisters, ages five to fourteen, I am watching our favorite TV show, "Touched by an Angel." We've been having trouble concentrating because the voices and laughter in the kitchen are so loud. We shake our heads no. Before my father can call to us, we are walking into the kitchen.

Our grandfather, Luther Angell, is sitting at the table. His three sons and three daughters-in-law surround him; some are sitting while others lean on a wall or the counters. Luther's wife, Joan, pronounced "JoAnn", stands by the stove with her hands on her hips. As I climb into my dad's lap, I wonder why Joan looks upset.

Luther begins his story in a dramatic, booming voice, "All

my life, I have had two goals. Now, there ain't nothin' to them and there's no rhyme or reason. But, I wanted to do these two things before I died."

Luther is a big man with a huge belly. He lifts up two fingers, one to represent each goal.

"My first goal was to go to Ash Grove, Missouri," he says, eyes twinkling.

Luther pronounces Missouri like older men, "Missour-uh" rather than "Missour-ee".

"Don't ask me why, I just wanted to go. My second goal..."

He pauses, looking around the kitchen at his audience, "...my second goal was to ride in a car down the main street of Centralia – naked!"

He shouts the word "naked" and I watch Luther's crowd laugh. He has drawn us all into his story. The TV show's drama seems bland in comParison. Surprisingly, no one seems to notice that our grandpa is talking to his young grandchildren about a crime of public indecency. My aunts, uncles and parents laugh as he shouts, "Naked!"

Joan is the heroine in many of Luther's tall tales.

Our grandmother, Joan, now has to hear the story retold to her grandchildren. She seems tired of the story; she turns her back on the crowd and begins washing dishes. Her sink faces a window to the backyard.

"Now, you might think that these goals seem a little silly and maybe they are. But, I bet you can't guess which one I've actually completed." Luther looks toward me, giving me a moment to wonder. I shrug my shoulders, trying to turn his attention back to the group and away from me.

"I'll give you a little hint," he says, grinning. "I am sixty-seven years old...and I have never been to Ash Grove, Missouri!"

Everyone laughs again, especially my father and my uncle, Jon. Luther has set the stage for one of his classic stories, priming the audience with two or three good one-liners. My father is laughing so hard I shake while sitting on his lap. Jon tips back in his chair and leans toward me.

Jon asks, "Do you know what that means? He's never been to Ash Grove...but...he's been down the main street of Centralia...naked!"

My eyes grow wide. I was too young to catch the joke the first time around. I am shocked, unsure of what to say.

"A couple years ago, there was a big storm in Centralia in the summer time." Luther drops his voice lower, making the storm seem scary. "After the storm passed, it was a nice, warm evening. I said to Joan, 'Come on, hun, let's go to town and see if anything bad happened.' I left the house in my old, striped nightshirt and house shoes. Joan was smart enough to leave her slippers behind, since my rusty old feeding pickup tended to have a layer of muck on the floorboards. She jumped up in Ol' Red wearing her gardening shoes and cotton nightgown."

When he really gets a-goin' on a story, Luther yells for emphasis. Sometimes those who sit too close to Luther are accidently showered in slobbers during the excitement of the telling. If he were a dog, he'd be the "wet mouth" kind.

"Then, we drove into town," he continues. "As we turned down Jefferson Street, I realized that the storm had knocked out the power for the streetlights. We cruised around the square and that area was dark, too. There were a few branches down, but other than that, the whole town seemed to be asleep. I guess no one else would be crazy enough to be out this late—except me and Joan! I was pretty certain that no one else could see us from their houses, no other cars had passed us, and the whole square was dark. I was sixty-some years old at the time; I couldn't help but wonder…will I ever have this opportunity again? I said to myself, 'Hell, this is the time to do it!'"

Naked on Main Street. A sketch by Luther's youngest son, Jon Angell.

I look around to see if anyone else noticed that Luther had cursed. None of our parents seem to notice; I decide it must be okay to curse while telling funny stories.

"Without hesitating," he says, "I pulled my nightshirt off over my head and stuffed it down by the floorboard. This was the moment to fulfill one of my life goals, but I had no intention of doing this alone. So, I reached over and quickly pulled Joan's cotton nightgown up and over her head. Before she could object, I threw it out the window into the warm summer air."

His hands wave as he motions dramatically, as if he is tossing the nightgown all over again. I giggle quietly and think to myself, "My grandparents were naked on Main Street?"

Luther continues, "I'm fairly certain the nightgown landed on one of the fallen branches. But then again, I'm not sure, because it was too dark to see behind us! I cruised the old pickup away from the city square and pulled onto the highway heading home. Joan knew about my life goals because she often heard me talk about them over the years. Yet, as I acted on them, she was shocked to be included. Feeling unprepared, she yelled, 'Luther, what are we going to do if we get pulled over?'"

Luther looks across the kitchen to his wife of fifty-some years and asks her, "Do you remember that, hun? We were both in our sixties and buck-naked! You were worried about getting pulled over by a policeman."

Joan turns and smiles mischievously, then continues washing dishes without a word. Luther turns his attention back to his audience, "You know what I told Joan? I said, 'Well, Honey, I don't know what you are going to do, but I'll just put my nightshirt back on!' And with that, I reached down to the floorboard and grabbed my nightshirt."

Jon leans back in his chair and laughs again. He laughs as if he has never heard the story before. In reality, this is the second time in one evening. Jon repeats the punch line for me, making sure I grasp the mature jokes. Jon emphasizes the punch line, "I don't know what you're gonna do, but I'll just put my nightshirt back on!"

This was the first time I remember hearing one of my grandfather Luther's theatrical, ridiculous stories. I was about ten years old. Through high school and college, I heard Luther tell that story a dozen more times. I always laugh at every punch line along the way; Jon doesn't need to help me understand the jokes anymore.

That first night though, I remember my grandmother Joan challenging Luther. She turned from the sink, dish cloth in hand. She scolded her husband, "Now Luther, you've told that story so many times, I think you actually believe it!"

Since then, I've been forced to wonder – was Luther's tale true? After all, Joan was the only witness.

When writing down Luther stories, I struggled between the role of journalist and storyteller. I wanted to question and prove every detail. When I went snooping for proof of Luther's tales, I found the shot put records and professional jersey number of his larger-than-life character "Big Ugly." Then, I ate lunch at Jack's Coronado, just up the road from the Columbia Livestock Auction. The tablecloths were white, just as Luther said.

Another time, my pastor told me about when she served at a church near Mount Mitchell where Luther was stationed in the military. She said, "The people there were the nicest. They were on their own time schedule, no one ever rushed."

I even checked the snowfall records for Alaska during the time Luther was stationed there for the military; the winter of 1954-1955 was a miserable one. In fact, until 2012, in nearby Anchorage this winter held the record for most snowfall at 132.6 inches – nearly double the average annual snowfall for that region.

Eventually, I stopped trying to prove the facts in each story; I found enough proof. Some of these tales might be big fish stories – the kind that grow an inch with each retelling. In stringing these stories together, which were told to me in starts and stops throughout my life, I created a logical time-line. Admittedly, this process has helped expand the big fish tales yet again.

Within these pages, fact and fiction are likely blurred at times. We will never know where the line is located. I've come to think of Luther's tales as a family folklore. I'll simply tell the stories to you, as they were told to me.

1920s
L.W. AND LOUISE ANGELL

CHAPTER 1

WISMAN WOMEN

"Have you heard the one about Louise's hat?" asks Luther.

I am sitting on top of an old Warm Morning style gas heating stove nestled in the corner of the room. This room used to serve as the office of the old Centralia Livestock Auction. Now, it is a cattlemen's version of a boardroom. Luther sits in an old office chair in front of the DTN monitor. The DTN is a specialized type of computer that delivers crop and livestock prices as well as weather reports. Luther sips on a Coke, checking the fat cattle markets and weather between our conversations. He wears a size 4X shirt, his belly bulging in the front, making the shirt gape open between snaps.

I tell him that I haven't heard the story, but I have. I want to hear it again. He probably remembers telling it to me before, but we both cooperate. Luther leans forward in his chair, eagerly preparing. "Louise and L.W. met when she and her family moved to Centralia. Not long after, they started dating. Then, they decided they was gonna get married."

L.W. Angell and Louise Wisman were my great-grandparents and Luther's parents, but I never met them. They both passed away years before I was born. I'll refer to them by their given names, but everyone in the Angell family called L.W. "Pawee" or "Pa" and Louise "Honey." L.W. was born in 1910 and graduated from high school in 1928. Louise was born in 1911.

"They got onto the train in Centralia," Luther's voice drops to a whisper. "They were heading east to St. Charles! But, they did not tell anyone where they were going – not even their folks."

Louise carried a small, hard-shell suitcase with white-gloved hands. She had made up her mind. Her back was straight, head held high. "Tomorrow," she said to herself while boarding, "I'll be Mrs. L.W. Angell." As she sat down, she adjusted her knee-length pencil skirt and matching jacket. That morning, she had carefully selected her best outfit and finest hat; she may not have a wedding gown, but she would wear her best hat to St. Charles on this important day. Knowing that onlookers may be watching as the train headed out of town, she turned her face away from the window, using her wide hat brim to hide her face.

Luther says, "They might have gotten out of town unnoticed, but Louise had a sister who lived by the railroad. Her name was Esther. She was an old maid. She liked to sit on the porch of her house with her friends."

Older women living next to a railroad today might complain about the constant noise from the coal trains. Not Esther. In 1929, she got to watch passenger trains. Her home's location elevated her to the status of unofficial town herald. After all, she knew everything about everyone.

As the train passed slowly by Esther's house, she looked carefully into each window. "Oh my!" Esther said to her friend. "I declare, that looks just like my sister's hat. I just know that was her hat going by there on the train. I didn't know she had plans to leave town, maybe she had relatives to visit."

Luther says, "At the end of the weekend, L.W. and Louise came back to Centralia on the train. They made the announcement to their families. My parents had run off and eloped!"

Louise may have known her folks could not afford a fancy, expensive ceremony and reception. By eloping, she never had to burden them with the responsibility of funding the event. Their wedding date was October 21, 1929 – just a little more than a week before "Black Tuesday" and the stock market crash that triggered the beginning of the Great Depression.

Luther pauses his story, adjusting his worn, misshapen Stetson, revealing neatly parted gray hair. Unlike most cattlemen I know, Luther never wears blue jeans. For years, Luther wore the old style Khaki Mesquite brand of western pants. This classic pant was made for generations of cattlemen who frequented the stockyards and sale barns across the country. But, as is so often the case, Luther outlived the Niver Western Wear Company. Now, he wears a lesser substitute: Wrangler Riatas.

Luther smiles and says, "Them nosey ladies were pretty surprised when they heard about that marriage."

Louise's sister Esther said, "By gosh! It really was Louise's hat! Who would have ever known she was leaving town to get married. I can't believe that my friend who runs the ticket counter down at the train station didn't tell me. She never said a word, but I know she knew – she sold them the tickets!"

The office at the old Centralia Livestock Auction has been our family gathering place my entire life. L.W. Angell purchased it in 1956 or 1957 after the Centralia Livestock Auction went out of business. L.W. and his sons, Luther and Buddy, used the building as a hog buying station until 2000. While it served as a hog buying station, the old maid, Esther, worked here as the bookkeeper. Since 2000, the building has been used for sorting and loading cattle. My sisters and I kept our show pigs here before the county fair during one particularly hot summer.

The office serves as the Angell family boardroom. The walls are covered with knotty pine boards. At some point, the bottom six inches of the paneling got wet. When it dried, it was wrinkled and faded. Although aged, the wood makes the room feel cozy. Rejected couches line two of the walls and three or four chairs fill up the rest of the small spaces.

Despite humble appearances, the planning, bookkeeping and deal-making happens here. From this small office, three generations of Angell men and a fourth generation of Angell women have discussed the business of owning cattle, sheep, hogs and livestock auctions. They have bought contracts, hedged and sold more than a thousand loads of fat cattle from this room during the past four decades. They have decided to buy two livestock auctions, in Bowling Green and Boonville, after meetings in this small office. A mouse might have run across someone's manure-covered boot shortly after they made a six-digit decision.

Although the mood may have been serious at times, no one notices every time a cattleman shuffles into the women's restroom. Cleaning supplies and forgotten boxes fill the men's restroom. For decades, bulls and heifers alike have used the bathroom with the hand-painted sign reading "Ladies."

Luther looks up from the DTN monitor and asks me, "Have you heard the one about Louise's hatpin?"

I shake my head no. I think I have heard it before, but I don't recall the details. He frames Louise's stories around the elements of her wardrobe. First, he introduced her eloping by mentioning her hat. Now, he brings up the hatpin. The classic, outdated wardrobe items evoke nostalgia before he even begins the tale.

"Shortly after my folks were married, they began living with L.W.'s grandfather, Joseph 'Tom' Angell," Luther says. "He was a grouchy old man and Louise was responsible for caring for him. It was a tough job, because he was often cross."

My father has an old photo of Joseph Angell hanging on the wall in his living room. Joseph has a long salt-and-pepper beard and a solemn, grooved face. An old photo has no meaning without a story; Luther fills the photo with life. I agree with Luther, Joseph did look grouchy.

"One day," Luther says, "Louise got tired of getting hollered at by the old man, so she left."

Louise wasn't thinking about the day she boarded the train with light-hearted feelings of a new beginning with handsome L.W. Instead, she stomped across town fuming to herself, "I won't stay there even one more day – no, not one more minute – while he treats me that way. If L.W. wants to deal with him, he can! But, I've had quite enough!"

She opened the door to the home of her mother, Angie Chamberlain Wisman.

Louise announced, "Mama, I'm moving back home – I can't stand that old man any more!"

Angie wouldn't hear a word of these complaints. The German mother met her German daughter with equal defiance.

Angie said, "You put your coat back on! You're going to walk right back across town. You are a married woman now. This ain't your home no more!"

She picked up Louse's hat and hatpin and shoved the hat forcefully onto Louise's head. With the hat snugly back in place, Angie grabbed her daughter by the shoulders and turned her toward the door. Angie sent her daughter marching back toward her new life. Louise chose this life when she signed her name to the marriage certificate in St. Charles. She had no other choice but to walk back across town. By the time she reached home, it was dinner time. She pushed aside her emotions and fixed a meal for herself, L.W. and grandpa Joseph. The joyful feelings of certainty and hope that she felt on the train to St. Charles were being tested. The work of marriage had begun.

"Did I tell you about how L.W. got his nickname for Louise?" Luther asks me. I shake my head no, driving Luther into another story about his parents.

"One time, Louise and L.W. got into an argument," Luther begins, using a serious tone. The topic of grandfather Joseph or some other subject must have razzed Louise's German side again. She felt the blood rushing to her face as her anger simmered just below the surface. The July weather was hot, so just a screen kept buzzing flies at bay. The stifling heat crept into the house with only a few breezes for relief. L.W. said something that caused Louise's temper to boil over.

Luther says, "She grabbed the honey pot off the kitchen table and threw it at L.W. – he tried to duck and miss the flying dish, but he fell out the window!"

As L.W. crashed into the yard, Louise must have thought, "What have I done?" She rushed to the window to see if her husband was okay. L.W. stood up and brushed himself off. He held up the pot, with a grin.

"Look, Honey!" he said while standing up to brush grass and dust off his clothing, "The pot didn't even break!"

"Damn him," she muttered to herself. "He's always so happy about everything!"

The fall didn't hurt L.W. Luther includes no hints of those details in his retelling. Instead, Luther launches into a punch line. "After that day, L.W. always called Louise'Honey'. And, that's how she got her name!"

These old stories have turned into cinematic black-and-white films in my mind. Louise Wisman Angell and Angie Chamberlain Wisman are three and four generations older than me in the family tree. They surprise me with their actions. No wonder these stories turn into movie scenes in my mind. These were women who eloped by train, said to their daughters, "This ain't your home no more" and threw dishes at their husbands. Oh, the scandal!

Notice Luther's hearing aid; all of the Angells are notorious for having poor hearing very early in life. At a young age, Luther's father, L.W., was declared legally deaf and was not allowed to fight in World War II.

Justin Angell, Mike VanMaanen and Luther Angell discuss the cattle markets and local sports teams.

Luther sips a Coke and enjoys reading the markets.

There are no neat and tidy hanging files in these cabinets. Instead Luther simply stacks the papers and notebooks in piles. At least they are in chronological order.

CHAPTER 2

ON BANKING AND BANKERS

I meet Luther's younger brother at the hog barn office. Luther was born in 1934 and Charles "Buddy" Angell in 1938. Buddy and Luther are brothers, but they haven't aged in the same way. At times, Luther pats his large stomach and announces loudly, "Hell, if I knew I was gonna live this long, I would have taken better care of myself!"

Buddy, on the other hand, still walks several miles a week and maintains a slim "cowboy cut" physique. Their differences don't stop at physical build; Buddy's memory is pristine. He recalls names and dates with enviable skill.

Buddy once told me, "I don't know why I do this, but sometimes I can remember even the exact date something happened more than forty years ago. But I'm not like Luther. He might remember the funny thing a friend said that day or the color of someone's shirt. I could never remember those things."

Luther likes to tell the stories that draw a crowd and make people laugh. Names, dates and background usually aren't important in spinning a yarn that causes laughter. In fact, with some of his stories, it is safer to leave out the names! Combining the memories of the two brothers brings out a wonderfully entertaining and accurate set of stories.

"What was Louise like?" I ask Buddy. Surely, there was more to their mother's life than eloping and throwing dishes.

"Our mother was German," Buddy begins. "Outside of our home, she was very quiet and reserved, just like most Germans in that era. On Louise's mother's side of the family there were the Chamberlains. Her mother was Angie Chamberlain Wisman. Louise's grandfather was named Edgar Chamberlain. He would have been your great-great-great-grandpa. He worked in construction. He helped build several of the houses and sidewalks that are still in use in 2014. He also helped build the original Centralia City Hall building."

Louise begins to seem more like a regular woman and less like a scandalous heroine in an old film. At one time, the floor in this office was painted red. Like old cow trails leading to a water tank, the soles of cowboy boots have worn the red paint away. A dark path leads to the tall counter where Luther pays bills and keeps books on his feeder cattle. The high metal stools have scratched away their own dark spots. Other faded paths lead to the Coke machine, DTN and ladies' restroom.

Buddy continues, "On Louise's father's side of the family were the Wismans. Louise's father was a train engineer on the Wabash Railroad. His name was Louis Wisman. Every day, his route was from Moberly to St. Louis and back again. He used to throw candy and pennies out the window of the train car on his way through Centralia. All the kids would run alongside to pick up the goodies. Louis Wisman was a member of the railroad workers' union. For some reason, the union decided they were not going to allow him to make the

L.W. Angell, Jr., looks dapper and serious. Notice the pocket square and coordinating tie. This photo was likely taken after the untimely death of L.W.'s father in 1925.

round trip from Moberly to St. Louis anymore. The union said he had to stop in Centralia, just about thirty miles before he arrived home in Moberly. He decided it would be a big hassle to find a ride from Centralia to Moberly every afternoon. Instead, he moved his whole family to Centralia. That way when he got off of the train, he was home. I guess he figured it would be easier to get a ride from Centralia to Moberly every morning instead of every evening."

In addition to the hand-me-down couches, the office has an old-time Coke machine. During my childhood, Luther kept it filled. He turned off the payment function so we could grab a Coke – in a glass bottle – anytime we wanted. He kept the machine stocked until Coca-Cola stopped delivering returnable glass bottles sometime in the late 1990s or early 2000s. A couple of wooden crates full of dusty Coke bottles sit outside in the hallway, as if fifteen years later, Coca-Cola might change its mind and we'll begin ordering glass bottles again. The grandkids weren't the only ones who enjoyed a free Coke on Luther's tab. The Amish in the nearby community of Clark also enjoyed their square bale hauling trips to the hog barn, because a cold, free soda waited inside.

Luther's voice painted a picture of Louise's temper, confidence and defiance while Buddy added softer shades. Buddy explains her German heritage, a reserved personality and her father's work. Buddy reaches up into the family tree again to describe their father's life.

Buddy says, "When L.W. was fifteen years old, his father died suddenly in 1925. As the oldest boy in the family, it became his responsibility to provide for his mother and siblings in place of their father."

Buddy begins at the place of tragedy in L.W.'s life. He lost his father quite young, forever altering their reality. L.W. had dark, serious eyes. While in his twenties, he wore his thick, dark hair lightly oiled and parted. The gentle wave of his hair gave him a mature, handsome look. His nose and eyebrows were strong, almost oversized, and masculine.

14

Buddy continues, "His father was Luther Washington Angell, Senior. He had a small herd of dairy cows and worked as a mailman. After he died, the family was no longer getting the check for mail delivery. He left behind debt on a mortgage on their home and farm. L.W. dropped out of school to start working. He planned to milk the dairy cows and begin repaying the debt."

The small farm Buddy mentions is near the Centralia Intermediate School and the water tower. One day, the banker stopped by the farm to let L.W. know that they would have to sell the cows to make the next payment. L.W. was furious.

"Don't you know what this means?" he asked. "If we sell the cows, I'll have no way to make money for the next payments. I know I'm short this month, but that's because Dad's mail check is gone."

He pleaded with the banker, but the banker only saw a young boy where a man was trying to stand. Instead of giving L.W. a chance to step up and support his widowed mother, the banker forced the sale on the cows.

Within a few months, they lost the farm. L.W. said to his mother, "They never even gave me a chance to try. I'll never trust a bank again."

L.W. never forgot the loss of that little farm. It was hard on him. A few years later, L.W.'s mother remarried and moved to St. Louis with her new husband. L.W. went with her, but he hated the city. He came back to Centralia, lived with relatives and finished high school. After high school, he started working right away and got married to Louise soon after.

Buddy says, "His whole life, L.W. had a lot of resentment toward banks. You see, he was never the type to tell me and Luther what to do. When we started buying our own cattle and farms, he never told us if he thought we were making a mistake. I guess he just thought it was better for us to learn from experience. Except, one time, I remember he gave us

15

In a rare moment of quiet, Luther, Buddy and L.W. pause to sit in the Ol' Blue Room at the Columbia Livestock Auction.

advice on mortgages."

L.W. was proud that his two sons were working hard. He never told them he was proud, but he hoped they knew. He watched as the boys bought their first farm together, resisting the urge to put up any collateral or offer any advice. Finally, L.W. gave in.

"Boys, you are on your own now and you can do what you want." L.W.'s strong nose and heavy eyebrows matched his serious tone. "But, if I were you, I would never borrow so much money that the bank starts telling you what to do. Don't ever give a mortgage on anything. If you give a mortgage, then the banker becomes your boss. He'll tell you what to do and when to do it. If you can live with that, go ahead and give a mortgage. But I can't live with that."

Luther and Buddy realized the importance of that moment. L.W. never gave advice, so when he spoke, they listened. He didn't mention the story of his own father's land at the time, but both Buddy and Luther knew that was at the root of L.W.'s advice. In short, he could have said, "Don't do what my father did."

"Another thing," L.W. added. "If you already owe so much money that the bank can't lend you any more without a mortgage, take that as a sign. Maybe you already owe enough."

Buddy says, "Luther and I listened to L.W. and we never gave a mortgage on our cattle. To make things work, we used two banks. We banked in Centralia and Sturgeon because both banks had pretty low lending limits. Since we had access to two banks, we could make it work. Our bankers sometimes said, 'Well, you are at your limit. You'll have to mortgage this farm if you want to buy those cattle.' When the bankers told us that, we didn't buy the cattle."

L.W. continued to resent banks throughout his life. At one point in the 1970s, L.W. and a few similar-minded friends applied for a bank charter. They were used to old-fashioned banking. A young, new banker from Kansas

City moved to Centralia and had lots of policies and procedures that didn't go over well with the farmers and cattlemen. For example, the farmers were used to walking in, explaining a project, and either being accepted or denied for the loan almost instantly – no paperwork necessary. The new banker wouldn't lend money without holding a monthly meeting and talking with the board of directors.

Buddy says, "L.W. and a few others decided to set up a new bank. They created a board of directors, collected the necessary money and hired a lawyer to handle the charter. The lawyer was absolutely certain the state would accept the proposal. However, they rejected it. The board of directors returned all the investment money when they realized that they weren't going to start a bank."

Many farmers and cattlemen can swap stories on banking nightmares; the relationship between agriculture and banking can be a tenuous one. What cattlemen and farmers need is a banker who will stand by them when the prices drop – not if they drop, but when they drop. They want a banker who will, in a long drought, overlook the percentages and the debt-to-equity ratios. They need a banker who will look them in the eye and say, "We'll get you through to next season – just hang on. You've done it before; I know you can do it again. Go home and work."

I thank Buddy for his time and we finish talking about family history. I never question the details of Buddy's stories. He is far too serious to exaggerate and carry on the way his wild brother Luther does. I trust each detail Buddy shares as fact. As he drives away, I can't help but notice how much cleaner his pickup is than Luther's.

The two brothers couldn't be more different. From the way they speak to their physical build and trucks, they are stark opposites. Buddy selected a modest two-door Chevrolet truck. It is very clean and dent-free. Luther chose to buy a cherry red Avalanche. A few years later, when Luther found out Chevrolet was going to stop making Avalanches, he traded in for a new one. This time, he picked a silver model with chrome handles and wheels. To Luther, a vehicle should – above all – be fun! Due to Luther's chronic poor driving, his trucks are always full of dents, dings and scratches.

Although they have different tastes, both would provide similar advice on purchasing vehicles, a direct result of L.W.'s advice to them on farm mortgages. L.W. would have said, "If you can't pay cash, you don't need the new vehicle." The Angell brothers are cautious about getting upside down on an item, like a vehicle, that loses value over time. To them, interest-bearing loans should be reserved for items that appreciate in value. In this family, that means land, cattle or other livestock. Taking out a loan on equipment, houses or cars is something L.W. would have been leery of.

Times have changed; farms, homes and especially vehicles are priced differently in 2014 than they were in 1940. However, the whole philosophy on treating loans and mortgages conservatively is rooted in L.W.'s childhood tragedy. It takes a long time to forget losing your home, farm and cows. The experience was painful enough that three generations later, his great-grandchildren try not to take out loans on cars.

Angie Wisman holds her grandson, Luther, in 1934. In the background note the A.B. Chance factory located in downtown Centralia.

As a young boy Luther took a fancy to his grandmother's green rose-patterned dishes. He always asked to play with them, but she had to say no. When Angie passed away, Luther inherited his favorite dishes. Apparently, the Wild Man has good taste in whiskey, cattle and china patterns!

CHAPTER 3

EYES FOR OPPORTUNITY

I meet Luther at the office of the hog barn again. He glances at the monitor of the DTN. The screen awakens with a burst of activity. Price quotes for the start of the morning's trading of the commodity futures in Chicago start feeding into his machine. Luther's attention is directed to what farmers and ranchers all refer to as "The Open." DTNs have an outdated 1990s look, with chunky monitors and no keyboards. Four arrow-shaped keys and few others provide the navigation buttons, perfect for hands still covered by leather gloves or dirty from work. I've never visited a livestock auction that didn't have a DTN.

Even though he is eighty, Luther still feeds out several hundred cattle each year. The prices are strong lately, but a seasonal decline is expected each summer, so everyone must decide if they are going to hedge their cattle or risk staying on the open market.

A majority of the Angell men are in the cattle business – Luther, Luther's two sons Jon and Justin, Buddy and his son-in-law Mike. They run four to six hundred-pound calves on their grass farms. They don't row crop farm one single acre; it is all pasture, a rare choice in this region. They will grow the cattle on grass to about eight hundred pounds and then ship them to a feedlot in western Kansas. They also run two livestock auctions in central Missouri.

"The Open" must be uneventful with no dramatic price movements, because he looks away from the monitor and back toward me.

Luther says, "Well, I think I'll tell you about the three places we lived growing up."

It delights me to know that he has planned in advance which stories he will tell today. I open my notebook and turn on my recorder. Sometimes, I've felt that our family stories were my personal responsibility. If they were lost, it would be my fault. Every time the light on my recorder turns red the burden lightens, I have captured one more vivid story from time's slow decay.

Luther says, "After L.W. and Louise eloped, they lived on a few acres and had a small herd of dairy cows. They lived in Centralia on the southeast corner of Jefferson Street. In 2014, this is the site of the Chester Boren Middle School track and football field. They had three children: I was born in 1934, Buddy in 1938 and Rosemary in 1940. While we lived next to the railroad, the coal trains stopped and unloaded near our house. When the railroad workers unloaded the cars, a few pieces of coal fell off of their shovels. The busy men did not have time to pick up the pieces that fell to the ground. L.W. made it our job to walk up and down the tracks picking up the forgotten pieces of coal. We heated our house all winter long with that coal. I'm not sure what made L.W. happier, coming home to a warm house or knowing he found a way to heat it for free. During the summertime, the railroad did not mow the grass, so we took advantage of that, too. L.W. hired a little black boy to walk our gentle dairy cows up and down the grass strips beside the railroad to graze from the end of a rope."

Luther's choice of the phrase "a little black boy" represents the era he grew up in. This was the politically correct terminology in the 1930s and 1940s. By now, he should have modernized and said "our African American neighbor" or simply "our neighbor."

It reminds me of Luther's DTN. A younger man might check "The Open" on his smart phone using an app from the CME Group, but Luther still uses the DTN. At eighty, his choice of words is the same as his choice of technology: outdated. He means no harm; he just hasn't kept up with the times.

Their life on the little farm place next to the railroad was pleasant. What more could a cattleman want? They had free grass in the summer and free coal in the winter. Luther relays these stories with happiness. He jokes about fact that they heated their house and fed their cattle by scavenging. He doesn't highlight the hardship, only the resourcefulness.

He continues talking, "Now, all this was before L.W. got into the livestock auction business. At this time, we just had those few dairy cows."

Even without the bank's support, L.W. eventually picked up right where his father left off when he died unexpectedly.

Luther says, "L.W. sold the milk and cream around Centralia. That wasn't enough to make a living, so he did odd jobs. He was a referee at ball games, too."

Anybody can be a referee at a ball game, but L.W. got the idea to be the concession man at halftime, too. He figured rather than sitting around waiting, he could be out in the crowd selling candy and cigarettes. When the buzzer rang, he set down the variety of merchandise he was peddling and went right back out to the court to continue refereeing the game. L.W. took advantage of the resources available to him, like grazing his cows along the railroad and selling candy at halftime. Anyone could have done those simple actions, but it took some ambition on L.W.'s part.

Luther says, "L.W. figured out a way to make a little money off of his knowledge of dairy cows. L.W. realized that he wasn't the only dairy farmer in the area with dry cows. At this time, many of the families in Centralia kept a family milk cow in a small barn or shed in their backyard. Even the families who lived in town often had a milk cow. With one family cow, it made no sense to keep a bull. Instead, folks milked their cow for several months and then traded for a new cow when the other dried up. This made for a chance at a steady stream of business for L.W. He began buying, selling and trading dairy cows, but always just one cow at a time. It wasn't long before L.W. had a constant flow of customers. Now, he needed a volume of dairy cows to supply fresh cows to his many friends and extended neighbors around town. About this time, folks started using thirty-two-foot trailers to haul livestock, which helped L.W. cheapen the freight on the volume of cows he could now trade. So, he started going down to the sale in Joplin - a five-hour drive - on Fridays. He bought a load of fresh, little Jersey milk cows and hauled them back home to Centralia. Anyone who needed a new milk cow called L.W. He would sell the fresh cows and buy the old, fat cows that folks were looking to get rid of."

I reply, "So that was sort of the beginning of the Angells buying and selling cattle?"

"Ya, I guess so," Luther says, "I think he just figured most folks didn't have time to make the trip to buy the cows. And, he liked to do business."

From a young age, L.W. had no father to rely on for advice or assistance. What did he have to lose by selling candy and cigarettes at halftime? What did he have to lose by trading on dairy cows? He had already lost his father, their farm, his boyhood home and their first cows. Essentially, he had nothing to lose.

While the rest of the country reeled over the Great Depression, L.W. seemed to be recovering from his life's first tragedy. After all, the Great Depression can't hurt you if you've already got nothing. A dramatic drop in stock prices on Wall Street isn't very problematic if you don't have any money.

What L.W. did have was a new wife, a growing family and everything to gain. He banked conservatively, took risks and perfected his cattle swapping skills. Because of his opportunity-driven worldview, he became the patriarch of our family. Throughout his life, L.W. continued to find and develop new businesses.

Trading milk cows was just the beginning. He joined a livestock auction, began buying cattle at the Kansas City Stockyards, opened a hog buying station and started a western store. He also bought and sold thousands of cattle, sheep and hogs. Today, the second, third and fourth generations below L.W. continue on the same paths he pioneered.

CHAPTER 4

L.C. HENDERSON

Buddy and I meet at the hog barn office. He has remembered a few stories about L.W. that I might enjoy. Buddy's granddaughter, Erin Gibbons, and I ran track together throughout school. She was one grade ahead of me and we ran on relay teams together. When I was in seventh grade and she was in eighth, we set a record for the 4x400 meter relay. In high school, we ran together again for three more years on the 4x800 relay team.

Buddy and Luther enjoy watching track and field events. They rarely missed a chance to watch any of their grandchildren play sports. They often arrived at the meet in cowboy boots with the faint smell of a sale barn lingering on their clothing. Luther stood by the final turn where he could cheer for us during the last, and hardest, two hundred meters of the half-mile race. When I passed, he would yell wildly, "Run like the wind, Angell! Run like the wind, Angell!"

For Buddy and Luther, tracking our relay times was similar to tracking the cattle market. They both were excellent at remembering our splits, previous best times and records for each event. Just as they could recall the prices on each class of cattle sold in a sale, Luther and Buddy knew every time and placing at the meets.

Our two-mile relay team eventually broke the school record with a time of 9:48. We placed third at the state track meet. Before we stepped onto the podium, I completed my usual post-race ceremonies by vomiting in the grass.

The previous record time was set in 1980. This was the first year that track races were measured in meters rather than yards. By chance, Lori (Angell) VanMaanen, Erin's mom and Buddy's daughter, was on the previous record-setting relay team. My father ran the half-mile and medaled at state on a 4x800 relay team, too. The 800-meter race really does run in the family!

Although I've graduated, Buddy and I visit about the track teams and local sports, but eventually we work back in time toward L.W., the great-grandfather I never met.

Buddy says, "Things were so different back then, it is a little hard to explain. But, I was thinking…did you ever hear of L.C. Henderson?"

I shake my head no.

"Well, L.C. was one of those old-time Centralia characters," Buddy explains, while settling in to a chair. "There really aren't many left like him. Luther is about the closest thing to the men like L.C. or L.W. Those men, they just liked to deal with one another. Do you know what I mean?"

"I don't think so," I reply.

He continues, "L.C. and L.W. were real good friends. They were always trying to sell one another something. They just liked to deal with one another. L.W. dealt in land, cattle, sheep, hogs and western clothes. L.C. Henderson didn't have any livestock, but he had a gas station, a furniture store and

29

he dealt in real estate. L.C. and L.W. were always trying to sell the other one something. They would yell at each other and argue over prices. Sometimes I thought they were going to come to blows. But that was just the way they did things. You see? They liked to deal with one another."

I ask, "Like all of the pocket watches L.W. bought?"

"Exactly," Buddy says, referring to the hundreds of old-time pocket watches that L.W. bought over the years. A huge glass case features the ones that L.W. passed onto Luther. There are probably a hundred in the case, which means his original collection divided between the three kids had at least three hundred. He enjoyed buying them because he thought they might appreciate in value.

Buddy continues, "L.W. and L.C. had the same philosophy, so they liked doing business together. They believed a fellow should 'buy low and sell high'. If a person got out a plat book and looked at all the names on the land in a ten-mile circle around Centralia, I bet you their names are on half of the land at one time or another. L.C. was the real estate agent in the area. He was always coming to L.W. with deals. So they

Luther's portion of L.W.'s pocket watch collection on display in his home.

bought and sold lots of properties together from the 1940s to the 1960s. They never held onto anything for very long though."

Land values have risen so sharply in the past decade that a trader's mentality on land is completely obsolete for my time. The days of trading farms or agricultural real estate as if they were fresh or dry dairy cows are gone. In my lifetime, I expect that many of the farms around Centralia will only change hands one, maybe two times. The majority of the land will be passed down from generation to generation. If things go well in an estate plan, family land is never offered for public auction.

"Now," Buddy continues, "sometimes they weren't partners and they just sold each other things. One time, after the Second World War, L.C. had access to army surplus items. For a while, he dealt in old army Jeeps. He never had very many, but he'd buy one or two at a time. After he sold those, he'd get a couple more. Somehow, he ended up selling one to L.W. Me and Luther thought this was pretty funny. Our father had absolutely no use for a Jeep! I mean there was not even one practical reason he should have bought that Jeep."

"But it didn't matter," I say, "because he had fun buying it and he'd have fun selling it, too!"

"Exactly," Buddy says, "he ended up selling that Jeep within a year or two. I don't even know who he sold it to. You know, looking back now, it wasn't such a bad purchase. At the time, Jeeps were the only four-wheel drive vehicles on the market. We never used horses and our two-wheel drive pickups were always getting stuck in the mud when we tried to get cattle up. That Jeep would have made a great farm vehicle. Too bad none of us ever thought about that. We should have bought a couple more Jeeps off ol' L.C. Henderson."

WORLD WAR II: 1939-1945
CHILDHOOD OF SCARCITY

CHAPTER 5

BAD NEIGHBORS

Luther says, "A few years later, we sold that little farm place and moved north up the road. The new farm was sixteen acres. Today, the Centralia McDonald's drive-thru and parking lot sit where our barns and house used to stand. It was a rundown farm; the ramshackle barns were just enough to get by. Everything was falling down and none of the gates or fences matched. Let's just say, it was no show place!"

He grins; it delights him to think a funny lookin' farm used to sit in the spot where folks now order their cheeseburgers and fries.

LEFT: Luther, Buddy and Rosemary were born in 1934, 1938 and 1940, respectively. For many years, the three siblings held a family reunion each summer over the Fourth of July weekend. Seated left to right: Charles "Buddy" Angell and wife, Sherry (Wilson) Angell; Rosemary (Angell) Boender; Joan (Gassett) Angell and Luther Angell.

His voice booms, "Every kind of livestock imaginable roamed the farm at one time or another! We had dairy cows, hogs, goats, chickens, a couple horses and a mean old rooster."

He ticks off the long list of livestock, counting them out on his fingers. He remembers them like old neighbors.

He says, "Our laying hens were quite a disorganized bunch. We built little roosting boxes for them to have nests. We hoped they would peacefully lay their eggs, but they did not use the boxes. It was like a year-round Easter egg hunt with them old gals."

As Luther describes the chickens, it almost sounds to me like he is describing the women of a church choir or the local gardening club. He is still fond of these farm animals.

He smiles again saying, "One time L.W. moved an old tractor and we found a dozen eggs laying all around it! No telling how long they were laying there before we found them. The eggs were all spoiled and we had to throw them out. I don't know why they didn't just use the nice roosting boxes we built for them!"

Luther was older at the second home, his memories are more descriptive and he's begun helping with the family chores. Did these chickens mark the beginning of Luther's legendary loathing and disdain for chickens and all things fowl?

"Every day, our rooster attacked me!" Luther says angrily. "I swear he waited for me to come around the corner of the barn. Then, he pecked at my arms and legs until I dropped the bucket of feed. The grain would spill all over and he would fight me for it while I tried gathering it up off the ground. It seems like all roosters get mean when they are old."

Luther leans back in his chair and adjusts his gray Stetson, remembering the barnyard bully. I'm sure now that it was the roosters and chickens at the farm that set Luther on the path of becoming an anti-chicken advocate.

"Pound for pound, I bet me and that ol' rooster were equal there for a few years." He says seriously, "I was sure tired of getting beat up on. One day, I was ready for him! He came around the corner right at me and I whacked him upside the head with a bucket. I did that every day for about a week. Then, that mean 'ol rooster finally left me alone! We had an awful lot of hogs, too. Most people at that time still raised hogs. Them damn hogs was always getting out of their pens! Pigs are always rooting around and messing stuff up. This was kind of a problem since the Centralia city cemetery was our closest neighbor."

Luther liked the chaos of that tiny, rundown farm. The little barnyard was mostly brown dirt with a few patches of dry grass; animals and three little kids overran the small yard.

Luther says, "I wish we could have had some old lady with a big garden for a neighbor. It would have been much easier to deal with her anger over lost tomatoes than half the town. Everyone got pretty agitated when our hogs got out and started rooting up their relatives."

This was frowned upon; no one wants a hog to be rooting up the freshly-filled graves. When holding a graveside service, dodging piles of brown pig poo should not be the first thing on a person's mind! They should be mourning the newly departed, not hopping over piles of hog crap. Or, worse yet, searching for a fresh patch of grass to wipe a shoe. If the hogs embarrassed Luther, I'm sure it was even more stressful for his mother. Did she look out from her kitchen window horrified to see that the hogs were out again? Then, did she round up the kids and do her best to collect the wayward hogs?

Luther says earnestly, "We did our best to respect the cemetery; after the hogs got out we always gathered them up and put them back in as soon as we could. Course when we got wind that there was gonna be another funeral, we'd head over quick and crawl around on our hands and knees trying to smash all the rooted up clumps of grass back into place before the service began. Now, that wasn't too bad of a job for us kids, but it was pretty embarrassing."

If the chickens were the nosy, old women of the farm and the rooster was the barnyard bully, then the hogs were the rowdy teenagers.

"We also continued milking several dairy cows." Luther says, "We ran the milk through a separator out on the back porch. We took the cream to town to sell and the leftover milk was fed to the hogs. It was a common practice for the families in Centralia to drive into town on Saturdays. The small town would be abuzz with folks milling around visiting, shopping, selling or buying eggs and cream. After the shopping was done, a family might go see a matinee at the local movie house on the main street of town. One Saturday, we got home from town late in the afternoon. Right away, I saw that them darn hogs were out. Except this time, instead of heading over to the cemetery they busted straight on through screen door and onto our back porch! I guess they must have smelled some spilled milk from the separator. It was summer, so the windows were left open as well as the door into the house. What a mess!"

Luther's voice rises excitedly as he remembers their house full of hogs.

"We had hogs on the porch," he says, pointing as if a hog is underfoot. "We had hogs in the kitchen and hogs in the living room. They were just tearing up and having the run of the whole place by the time we got home! Our house looked worse than them rooted up graves."

"What did Louise do?" I ask, feeling awful for his mother.

"I think that was the maddest I ever saw her," Luther says. "Cleaning up after the hogs in the cemetery was easy, but putting our whole house back together again, cleaning up the hog poop and calming down our mama was a tough chore for everyone. Seeing Louise angry made the cemetery folks look mild in comParison. After a few weeks, the smell went away. She never did forgive the hogs for rooting up her house."

"I bet," I say. "I would have been upset, too. I wonder how many people who drive up to the McDonald's drive-thru realize all the excitement that used to take place there? It is ironic. Back in the 1940s your family was raising hogs, dairy cows and hens on that little piece of ground. Now, they serve bacon, egg and cheese sandwiches out of a drive-thru window."

Luther says, "I bet they have no idea it used to be a farm, but I was sure ready to move away from that cemetery. A McDonald's makes a much better neighbor to the cemetery than our farm did. Those were the first two places we lived. Then, in 1948, L.W. bought a farm east of town on Highway 22. That's where I live today."

After dealing with upset neighbors for several years, Luther, Buddy and Rosemary were looking forward to some freedom from life on the edge of town. The quiet, country life – no more dealing with the demands of city living.

Luther's eyes turn gloomy and his gray eyebrows come together, "Boy, was I disappointed when I learned about our new neighbor. The new farm bordered the Centralia Country Club golf course. Here I was looking forward to a fellow farming neighbor and L.W. picked a farm place that bordered a golf course! What was he thinking? When the sows got out onto the golf course, we were in a world of hurt. You cannot believe how much damage a couple of sows can do to an irrigated, finely manicured golf green in a couple of hours!"

Luther and Joan live on the same farm today; they moved there after Louise and L.W. passed away. I'm used to seeing his cattle grazing near the golf course. One warm summer evening, my young cousins and I ventured outside of our grandparents' yard to play on the golf course. We played in the sand around the greens and found misplaced golf balls. It was great fun, but we must have been messier than we thought. Our grandparents got a phone call the next day about all the little footprints in the sand. We were unfamiliar with the rules of golf, we didn't know we were supposed to rake the "sand box" when we got done playing.

Luther says, "One summer night, our sows got out onto the eighth green. The next morning, there was no way L.W. could send me and Buddy out there crawling on our hands and knees to patch up and cover over that grass. It was hopeless!" His voice squeals and squeaks, because he is wound up and excited, "I think the country club's groundskeeper was even more worked up than the town folk after the hogs were digging up Uncle Henry's grave. The country club turned out to be our worst neighbor yet!"

I add, "I think each house got worse. At least at the first place you got free coal for the house and free grass for the cows."

Luther leans back smiling, "Well, now that I look back on it and tell you about our neighbors, I am starting to change my mind. After all, it was our livestock that trampled around the cemetery and golf course. Maybe we were the bad neighbors."

In 1948, L.W. purchased his farmstead from the Metropolitan Insurance Company.

CHAPTER 6

THEY WERE TERRIBLE

I knock on the door of my Great Aunt Rosemary's house. A stylish, tall woman greets me.

"You're here! Hello!" Rosemary exclaims. "You know you're the only one from Centralia to ever come visit me! I've been living here for years and nobody will drive up here. Come on in!"

I have an internship for the summer in South Dakota, but first I have to attend training in Minneapolis. I stop in Oskaloosa, Iowa to visit Rosemary and stay overnight at her house. Rosemary has short, stylish blonde hair and wears pink lipstick and bright colored clothing. She is Buddy and Luther's younger sister.

We sit down on the couch in her living room and she twists open the lid on a bottle of Diet Pepsi. "I'm terrible about drinking this soda." She says, "I order this stuff by the flat. I could order a whole pallet full I think. Can I get you anything?"

Rosemary might go out for supper wearing white capri pants, sequined silver shoes and a pink floral top – an outfit that most twenty-somethings would be thrilled to wear. Yet, here she sits, in her late sixties, looking stylish as ever. I hope I'm half as trendy as Rosemary when I'm her age.

Her voice rises and falls, much like her brother Luther's voice. Rosemary and Luther are clearly siblings. They both speak loudly, partly because they have become extremely hard of hearing like their father L.W. However, their bold personalities also cause their voices to raise higher and higher, arms and hands waving for emphasis.

"Did you hear what my son Tim did?" she asks, her voice reaching up to a higher octave. "He came here and took away my guns! He said I was too old for guns. Can you believe that?"

She takes another sip of her Diet Pepsi and smiles.

"I'm just a little old lady," she squeaks. "How am I going to defend myself? Tim said if someone broke in I would probably shoot myself first. I told him, he better watch out – I might go buy another gun. If I do, he better not come knockin' on my door in the middle of the night. We'll both end up in trouble."

She waves her hands, loading an imaginary gun, flashing pink manicured nails, bright rings and stacks of jangling bracelets on each wrist. I can't image taking a gun away from someone as lively as Rosemary; her sons are brave. I notice that her living room is decorated with antiques. Even though she dresses with the brightest trends, she decorates her home with a more reserved style.

I motion to the antiques and ask, "Do you like auctions or antique shops?"

"Ohhhhhh!" She exclaims mischievously, "I love auctions! Boy do I spend too much money sometimes. Auctions make you do that, don't they? One time, I was at an auction and an antique dealer was there. All day long, he kept buying everything I wanted! He was really getting on my nerves. I bid on dozens of items and I hadn't gotten a single one."

She tries to appear annoyed, but she still smiles and laughs.

"I like collecting antique children's toys," she explains. "When the auctioneer finally came to this little cart, I stuck my finger in the air."

She points to a small, tin toy, no more than eight inches long. Then, she reenacts her bid for the cart, sticking a long, pink fingernail into the air stubbornly.

"I wasn't going to let him get that cart," she says proudly. "So, I put my finger in the air and I didn't take it down. Do you know how much I paid for that cart?"

I say, "I have no idea."

"I paid two hundred dollars!" She says dramatically, "Now, isn't that the craziest thing you ever heard? But I wasn't going to put my finger down and let him get another thing."

She says "him" and the word drips with disdain.

"I wanted it," she announces, "so I bought it!"

We laugh about her purchase and I tell her about the time that I bought a gold ring with four emeralds at an estate auction outside of Centralia. I paid about $350 for the ring; at first I thought I'd made a huge mistake. I was a junior in high school and I should have saved the money for gas. I had the ring appraised for $3,800. I wish I'd had the nerve to buy the other three rings for sale that day, too.

"What was it like growing up with Luther and Buddy for brothers?" I ask as we settle back into Rosemary's fluffy floral couches. The three-hour drive from Centralia to Oskaloosa wore me out, so I'm glad to sit and visit. It is too bad her bothers, Buddy and Luther, don't visit. Her home and the antiques are beautiful.

Just like her son taking away the guns, she gets fired up at the thought of her childhood with two older brothers. Apparently, Buddy and Luther were so terrible to her that five decades did not cause her childhood memories of pesky brothers to fade.

"Luther was the worst," she begins telling me, "but sometimes he convinced Buddy to go along with his ideas. One time, Luther told me to climb the big tree out in our front yard. He said he would come up next. Once I was up in the tree, Luther told me to crawl across the branch leading onto the roof of our house. But once I got on the roof, he pulled out a saw and cut off the branch. He was just terrible like that. He left me stranded on the roof all afternoon."

I shake my head. What an awful trick!

Rosemary says, "Another time, our father, L.W., went away for the day and decided to leave each of us a little money, like an allowance. L.W. left the money with Luther and I didn't know anything about it. Luther decided to take advantage of that. Luther said, 'Rosie, I've got a deal for you. If you wash and wax my car, I will pay you.' I thought that sounded okay, so I spent all morning cleaning his car. When I finished, he paid me with my own money! Boy, was I mad when I found out about that!"

The emerald ring purchased at an auction.

Rosemary continues, "I think the worst prank they ever pulled on me was at the 'Fun House'. This was one of those times that Luther convinced Buddy to help him. They convinced me to be a paying customer at their 'Fun House' which was really just our old barn. When I went inside, they blindfolded me! They said I had to solve a mystery. Then, they stuck my hand in a bucket of oil. They teased, 'Guess what is inside the bucket, Rosemary!' I screamed, but of course that made them laugh more."

As she talks, I become thankful that I had sisters. I am the oldest daughter in a family of four girls. I remember playing horses or dolls but nothing with tar or pranks. We

Rosemary learned to be fun and tough growing up with two pesky older brothers.

45

sometimes fought, but never the antics Rosemary describes from her childhood with two older brothers.

"After the mystery bucket, they took me to the hayloft," she says while motioning higher into the air. "I was still blindfolded. Then, they dropped me out of a hole between two boards! I fell into a hay pile below. Can you imagine? Getting dropped while blindfolded? It was terrifying! I ran to the house in tears. I think our mother made them give me my admission money back."

"My goodness, they were terrible to you," I say. "Maybe you should be glad they don't come to visit you in Iowa after all!"

CHAPTER 7

THE MEXICO
LIVESTOCK AUCTION

Buddy says, "In 1942, L.W. became a partner in the livestock auction in Mexico, Missouri. The Mexico Livestock Auction had ten owners."

Mexico is located about twenty-five miles to the east of Centralia on Highway 22.

"This was a pretty common arrangement for a livestock auction partnership at the time, but it is not very common anymore," Buddy continues. "The barn was built new, just about the time he became a partner, on Morris Street in Mexico. They hired a painter to do a mural of a western scene in the sale arena. This arena and the western mural were sure pretty!"

Most people would not describe a stinky livestock auction as pretty or beautiful, but the men in the Angell family see things differently. They get downright nostalgic about the layout and design of a sale barn, reminiscing on the rooms, furniture, pens, "check-in shacks" and even the dark alleys leading to still more pens.

L.W. Angell gravitated to the livestock auction business. It suited him well, because he never found a job too dirty, stinky, cold, hot or hard. My father, Justin, likes to say, "There's no place in the world colder than the back of a sale barn in January, and there's no place in the world hotter than the back of a sale barn in July!" At times, the strong concentration of livestock urine causes an ammonia-like smell to permeate the air and burn the eyes. But, none of this bothered L.W.

Besides, the life of a farmer or rancher was too isolated for him. He loved to buy, sell, trade and swap with the men like L.C. Henderson. He enjoyed trading the dairy cows all around Centralia with his friends and neighbors.

Working alone on a farm day after day would have seemed boring to him. Instead, he wanted to get out and deal with folks. I can't think of a better place for buying and selling than a livestock auction. All day long, folks are buying and selling. And, what did L.W. do at the livestock auctions? He was an auctioneer – at the center of all the action!

"During that era," Buddy says, "they used handwritten tickets to keep track of the livestock. The clerk would put the date, description and seller's name and information on the ticket and then pass it on to the next guy. The next man wrote down the selling price and the buyer's name and information. The last man would imprint the ticket with the weight from the scale. He just had to grab the ticket and squeeze it. Then, it would imprint through the carbon."

Today, a clerk types the information into a computer and the information is transmitted wirelessly to another computer in the main office. There is still a paper trail on each set of cattle, but computers play a large part.

Buddy and Luther were four and eight years old when L.W. became a partner at the Mexico Livestock Auction.

Although they were young, they often rode along to the sales with L.W. because they were on Saturdays. School never interrupted their chance to spend a day at the sale barn.

"L.W. worked as an auctioneer part of the time," Buddy says. "Back then, most folks knew how to chant a little. As I got a little older, I decided I wanted to learn. It was a good place for me to learn to chant because I was exposed to lots of different styles. Each sale might have five or six auctioneers in one day. Each man sold a different class of livestock. The sale would start out with the first auctioneer and he would sell all the hogs. Then, another man would sell the dairy cows. A third auctioneer would sell the feeder cattle and a fourth sold the cows."

At a modern livestock auction, there might be two auctioneers who sell all day long. An auctioneer in the morning might sell the cull cows and cull bulls, then after dinner another man finishes the sale by selling all of the feeder cattle and fat cattle.

Buddy says, "The very best auctioneer I heard at the Mexico Livestock Auction on Morris Street was Eddy Buckner. He was the very first World Champion Auctioneer."

The Livestock Market Association held the first World Livestock Auctioneer Championship in June of 1963, almost twenty years after Buddy began listening to Eddy's chant in Mexico, Missouri. The first contest was held at the Cosmopolitan Hotel in Denver, Colorado. Eddy Buckner competed with twenty-three other contestants, but he came out the winner. Each man auctioned off the same twenty head of cattle over and over again.

The high-quality auctioneering had a strong impression on Buddy, and sometime shortly after the age of four, he decided he wanted to become an auctioneer, too.

"To learn to repeat some parts of the chants, I just listened and then gave it a shot." Buddy says, "Other parts were so fast I could not pick out the words, so I took a recorder along to the sales. I replayed the fastest parts over and over, until

I understood each word clearly. Back in the 1950s, all of the country kids went to the first eight grades in little community schools scattered all over the countryside. Every three or four miles outside of Centralia in each direction there was a country school. My mom hauled me all over the county to auction off pies. It was good practice for me."

Although Buddy desperately wanted to become an auctioneer like the men he saw at the Mexico Livestock Auction, he had to overcome his fear of crowds. The first time Louise took Buddy to a local pie auction, he took the stage and forgot everything he'd ever practiced. He acquired a bad case of stage fright, and another man had to begin the sale in young Buddy's place.

When Louise calmed him down, Buddy went back to try again. At first, he hesitated to look out in the crowd. But, he had to look for the bids, so he slowly raised his eyes and began chanting. The little schoolroom was crowded with many faces.

Buddy says, "I learned pretty quick who to look for in a crowd of potential buyers. The father or boyfriend of the young lady holding the pie was probably going to be the winning bidder."

The Mexico Livestock Auction was similar to many of the small, community-type barns in Missouri during the 1940s and 1950s. At that time, the Kansas City Stockyards still carried the most volume, so community sales were often small.

The unique thing about the Mexico and Centralia area was the number of excellent auctioneers that came out of the region. Eventually, the World Champion Auctioneer contest recognized several of the men who got started with their careers in this area. Eddy Buckner, Buddy's childhood favorite, won in the first contest in 1963.

In 1974, Ralph Wade of Greeley, Colorado was the winner. Although he was living in Greeley at the time, Ralph was raised near Fulton, Missouri. The next year, 1975, Ron Ball won and at that time he was living in Brush, Colorado. Ron was originally from Centralia. Ralph and Ron were younger than Buddy, but in later years Buddy recalled purchasing cattle that they auctioned off.

Buddy says, "Most people probably don't realize that three of the first ten World Champion Auctioneers were from central Missouri."

Of course, livestock auctioneers do more than say words really fast; the process is more of an art than a science. With time, a good auctioneer will memorize the names and faces of every buyer and seller. Additionally, he will also learn a tremendous amount about the inner workings of the market. He should know who each order buyer is buying for and the size of their weekly order. During a sale, the auctioneer should keep track of how "full" each order buyer is for the day. This way, he knows how high to expect bids and where to look for the bids.

What's more is that the auctioneer should also learn the personal dynamics of his market. Which order buyers get along well? Which ones sometimes compete? All of this

The Mexico Stockyards on Morris Street in Mexico, Missouri, circa 1940.

information helps the livestock auction maintain a competitive market. It represents good, fair business for both the customers and buyers.

In total, L.W. spent less than ten years at the Mexico Livestock Auction on Morris Street, but those years were important. He started auctioneering sales and got a taste of the sale barn business, and his young son, Buddy, also decided to become an auctioneer. It eventually closed down, but by that time he was a partner at the Columbia Livestock Auction in Columbia, Missouri.

CHAPTER 8

A CHILD'S WORK AND PLAY

I stop by my grandparents' house to hear a few more of
Luther's stories. My grandmother, Joan, fixes chili for our
supper. Rather than Grandma, her nine grandchildren always
called her Mamoo. The unusual name started because she
likes to wear long, flowing housedresses with bright, floral
prints. Somewhere in the South, this style is called a moo-
moo dress. The dresses emphasize Joan's short stature and
busty figure. She has a thick, but endearing, southern accent.
Her words are slow and languid, like a humid, summer day
in Georgia.

If I stop by at 11:00 a.m., I know I will find Joan still wearing her nightgown. Everyday, she sets her coffee on a small,
electric heating plate. She will still be sipping on the same
cup at 2:00 p.m. that she poured at 9:00 a.m., so it needs to
stay warm. She will spend all morning making a pot of chili,

tasting and adjusting the recipe along the way. Then, she can spend all afternoon simmering a pot of green beans with bacon and potatoes. At Christmas one year, she spent several hours standing over the stove slowly stirring, because she wanted to make homemade "drinking custard" laced with rum. Joan never rushes.

Although she is the grandmother I have always known, her son Jon remembers a younger version.

"When I was a kid, she worked all the time." He says, "She could get more work done in a day than two men could today. That might be an exaggeration, but not by much!"

"But still," I say, "I've never met anyone who can work at something as long as her."

Jon smiles, "That's because she has a slower sense of time and a compulsion to stay at a task for too long. Wouldn't you say she passed that onto me? And maybe you, too? How long have you been working on this book, again? Don't you make quilts?"

As I watch Joan taste the chili for the third time, I realize I am probably more like her than I realize. Jon and I may know different versions of Joan, but we both appreciate her patience, even if it borders on compulsion. Over the years, she made beautiful quilts and intricate, wooden doll houses, complete with electricity. She did all the electrical wiring herself. The perfect, uniform stitches on her intricate quilts and the extra tastes of each meal are worth the extra effort. The chili Joan serves is delicious. After supper, she clears the table and Luther and I go into the living room. He sits in his blue leather recliner.

I sit across the room in a floral wingback chair. I think Joan picked this chair up at an estate auction. Like Rosemary, Joan enjoys a good day of antiquing. Joan was with me the day I bought my emerald ring. She helped me keep bidding when I almost lost my nerve.

Their fireplace is unlit, but the mantel is lined with bronze statues of cowboys roping calves and riding beautiful horses.

At Christmas, Joan winds dark green garland between the
bronze horses. It is rustic and beautiful. Those frozen scenes
on the mantel represent a different time and place, probably
further west in the late 1880s. In central Missouri during
the 1940s, those rowdy cowboy days were replaced with a

Joan and one of her handmade doll houses.

55

One of Joan's beautiful quilts, made in 2007.

modern version that looked much different. Of course, there were still rowdy cattlemen, but their lives and actions fit their era. Luther begins to tell me more about his childhood as a farm kid.

He reclines and puts his hands over his round belly, full of chili, and says, "We had a real bad snow storm when I was about seven years old. It was the kind of storm that came up out of nowhere, so L.W. was unprepared. He still had cows out on fall pasture without enough feed or shelter when the snow began to fall. He pulled me out of school to help him move the cows to feed, water and shelter. L.W. picked me up in our junker truck and we drove out to get the cows up. We headed to a pasture he was renting north of Centralia. I did not expect to work outside that day, so all I had for shoes was flimsy rubber mud boots. They did not have any insulation and my toes were cold before we passed the city limit sign. I already wanted to go home and warm up."

Luther is a lifelong cattleman, but the stories from his life don't match with the bronze cowboy scenes on the mantel in his living room. His own father hated horses and never owned any for working or gathering cattle. L.W. would cheerfully buy a pony for his grandkids, but he never considered owning or riding a horse of his own.

Luther says, "There was no such thing as a trailer back then. Our plan was to walk the cows down the road to feed and shelter. We hoped to lead or follow the cows in our truck. Then, the damn thing froze up! It was cold enough that the lines froze. After that, L.W. decided we would just start walking. He called the cows and carried a couple flakes of hay; the hungry cows followed him through the driving snow."

Luther is warm and relaxed while he tells this story and looks as if he could doze off in a matter of minutes. Only his eyes show the emotions of a cold, frightened seven-year-old.

"L.W. left me to follow up the rear," Luther says. "I was scared to death."

Luther almost forgot about his cold toes as fear grew in his chest. The snow was blurring out the outline of his father's body. Then, suddenly, the snow swallowed L.W. up completely.

"I sure felt lonesome." Luther says, "I walked as fast as I could to keep up with them cows. I felt like I was trailing hundreds of cows across half the state. After a long time of tromping through the snow in my thin rubber boots, we made it to a barn. I thought my toes would never thaw out."

Luther sounds like a frightened young boy while he talks. Returning to manhood bravado he says, "In reality, it was only thirty or forty cows and a couple miles. No matter the distance, I'll never forget how damn cold I was during that blizzard! Seventy years later, every time we haul cattle in the winter, I'm real thankful for antifreeze, cattle trailers and insulated rubber boots. Hell, back then, none of that stuff was even invented yet! There were lots of times when me and L.W. did things that might surprise you. Another night, we was over at the Fulton Livestock Auction watching the sale. The sale got over late that evening and we got ready to leave in our same old junker truck. The engine started just fine, but the lights would not turn on. No matter how many times he cussed them, they didn't even flicker. We needed those lights; the drive from Fulton home to Centralia is about forty miles of curvy roads."

They were growing desperate. Their junker truck was the only one left in the sale barn parking lot. The longer they waited, the longer the night seemed.

Luther says, "L.W.'s solution was to set me out on the front fender clutching a flashlight. I should have been scared, I guess. But I wasn't. It didn't matter either way. L.W. would not let me be scared."

"I can't believe that!" I say to Luther, "Today, they have passed laws to keep people from riding in the bed of a pickup. I wonder what would happen now if a parent used their kid for a headlight?"

Luther seems indifferent to my surprise. To him, that was just what he had to do in order to get home that night.

He says, "I perched on the fender and shined the flashlight on the road. The dim beam had just enough light to keep us from going off into a ditch when we rounded a corner. We made it home just fine. Another time, a couple years later, L.W. had a straight truck with tall, wooden stock racks. One time, he bought a couple sheep and a cow at a sale and the end gate got jammed open. These gates were about four foot wide and they ran in a small metal track that allowed them to slide up and down. If the track ever got bent, the end gate could easily jam open or closed. Try as we might, our damn gate wouldn't come down. L.W. wanted to get home, so he decided I could stand in the end gate's place. I braced one foot on one side of the hole and the other one the other side. I reached, each hand high above my head, and grasped the sides tightly. I stood there all sprawled out, praying the animals would be real still all the way home. It was a long, scary ride through the night, not knowing if one of the animals was going to come out of the darkness kicking or butting me out onto the highway."

"Oh my gosh. That's awful!" I say. I am beginning to think that every kid born before 1940 had a terrible childhood.

"Well, it was a different time. Let me tell you about some of the games we played," Luther says, changing topics dramatically. He can tell I'm getting upset. "We worked all the time, but on Sunday afternoons, we got to play. Me, Buddy and our friends invented a war game on the farm. We called it 'corncob wars'."

He smiles, putting extra emphasis on the name of their game.

He says, "We devised a battleground, where five kids hid out in the barn and five others were the invaders. Both teams would battle! One team tried to defend the fort, while the other team attempted to overtake the stronghold. Our favorite part of the war game was our ammunition: corncobs.

To ready the ammo, we rolled the corncobs in fresh cow manure. It was a real fun game, until L.W. got upset. Our ammo caused cow poop to splatter all over the broad side of his barn. It wasn't a fancy lookin' barn to begin with and once we splattered cow manure all over the side, it looked a whole lot worse."

"Boys!" I tease, "Girls would never come up with a game like that!"

Luther nods his head, agreeing, and continues. "After L.W. put an end to corncob wars, we resorted to cow pasture football. The game was a lot like regular football, except it had one extra rule. A great defensive cow pasture football player always managed to tackle his opponent right into a fresh cow pie! It didn't really matter which team ended up with the most points, the real fun was tackling the other team's guys into a sloppy mess. My team usually won, because I was four years older than Buddy. All my friends were bigger than his, which gave us a bit of an advantage."

He adds proudly, "When we headed back to the house, his friends were always the greenest and dirtiest."

CHAPTER 9

THE CATMAN OF PARIS

Joan has been working in the kitchen and listening to Luther's stories. When she notices that Luther has paused, she hollers into the living room. She strains to make her voice loud enough for her hard of hearing husband. "Luther," she yells, "tell her the one about *The Catman of Paris.*"

Luther perks up at the mention and begins another theatrical reenactment. He just needed one small trigger and the whole memory of a boyhood Halloween night comes into focus.

He begins, "Every year, on Halloween, it was a tradition that the Vista Theater in Centralia played a scary movie. My friends and me wanted to see a hair-raising horror film to make our Halloween extra spooky." He darts his eyes back and forth as if someone is watching. He's trying his best to act scary to enhance his spooky Halloween tale. I play along.

He says, "The movie gave the grade school boys a chance to prove our bravery to each other. And, more importantly, to show off for the girls. In 1946, I was twelve years old. I was at the prime age for Halloween stunts and pranks. That year, my friends and I wanted to enjoy the annual scary movie. After watching the show we planned to top all the stunts and pranks."

Luther's childhood in Centralia was far different than mine. He had two movie theaters to choose from on the weekends. Our family drove twenty minutes to a neighboring town with a theater. The local theaters were closed by the time I was growing up.

He continues, "On Halloween in 1946, the Vista Theater played *The Catman of Paris*. The Catman was no doubt the villain of the movie. He looked like a vampire, with pale skin and long teeth. Except his teeth and skin weren't vampire features. They were catlike features. His fingernails were long and yellowish-brown just like a tomcat's claws. Worse than the teeth or the claws was the scary, "meeeeooooowwwww." He stalked around the city of Paris, searching for his next victim while letting out terrible "meeeoowwws" everywhere he went.

"Back then, our spooky night on the town only cost me and Buddy about a quarter apiece. The ticket into the Vista was only 'bout twelve cents. A small bag of popcorn cost a nickel and a bigger box cost a dime. After the movie, all my friends wanted to head uptown and get a milk shake. Some of my friends were inspired by the movie and were planning all kinds of cat-related pranks. I didn't let my friends know it, but I really didn't want to get a shake. That movie had been far scarier than what I expected. I wanted to go home! I was praying my mother would come to the theater and say I had to go home early. I was afraid that the Catman of Paris was out stalking the streets of Centralia—and I was going to be his next victim."

Luther continues to build the suspense, acting as if the

A movie poster for *The Catman of Paris*.

Catman could come into the house at any moment. He looks over his shoulder to the dim hallway before continuing.

"My prayers went unanswered," he says. "Or, at least they didn't bring Mom to the theater to save me. She left me no choice but to go along with my friends for a shake. I puffed up my chest bravely and said, 'Well, boys, that movie wasn't too bad. What's next?'"

Joan has almost finished cleaning up the kitchen. Just as her patience helped cook us a wonderful dinner, her un-rushed cleanup is impressive. She won't leave the kitchen until each leftover item is packaged in the perfect-sized Tupperware container.

Luther continues, "We headed uptown and along the way someone spotted a stray cat. 'Grab it!' my friends shouted. Someone grabbed the old cat and we carried it along with us. The stray cat did not like to be bothered by rowdy twelve year olds and it become rather rowdy itself. We tossed the angry cat into an unlocked car. We slammed the door and left the cat behind. Our Halloween prank was in motion, now we could go up town for shakes."

"You left the cat in the car?" I ask.

"Yes!" Luther says, "The next morning, those poor folks opened the door of their car and the cat flung itself out onto the sidewalk. It was gone before they even noticed what color it might be. I heard the inside of their car looked like a cougar had been trapped overnight. Every seat was shred-ded from the cat's frantic clawing. The vinyl seats, polyester carpet and fabric ceiling of the car had been scratched and clawed into pieces. Luckily, we never did get caught for that year's prank. After throwing the cat in the car, we went to get our milk shakes. I can still taste those shakes! They only cost ten or twelve cents. Mmmmmmmmmm! They were all good, except the one I had on Halloween in 1946, because the Catman of Paris was on my mind. I was thinking about the long walk home, hoping the street lamps would be brightly lit. After the shakes, Buddy and I started to walk home. We walked quickly and quietly. At the time, we still lived right night next to the cemetery!"

While the boys walked, the wind howled through the trees that lined the edge of the cemetery. A squirrel ran across the street and Luther almost jumped a foot into the air!

Luther's voice rises as he squeals, "Oh, what an awful neighbor on Halloween! Sometimes, when Buddy and I were feeling brave, we would cut through the cemetery together. But on Halloween night in 1946, there was no way I'd cut through that spooky place. We were going the long way around! Buddy and I stuck to the brightly lit streets and made it home. My nerves didn't settle for a long time after we were safe inside the house. I swore to never go to the Halloween movie again. Starting next year, I was gonna fake sick."

Joan's living room is so peaceful compared to this dramatic tale. Bookshelves line two walls; the east shelf holds all of Luther's books and the west shelf holds Joan's books and doll collection. Luther favors murder mysteries like Truman Capote's *In Cold Blood* and authors like Tom Clancy or Dan Brown. Apparently, he overcame his fear of scary movies and moved on to mystery books.

Luther says, "A few days after the movie in 1946, my mama sent me out to the barn to collect eggs. Some of them old chickens liked to lay them way up high in the back of the hayloft. I went up to get them, carefully dodging the spider webs and cracks between the bales of hay along the way. I made it to the back of the loft, where only a few slits of light came in between the barn wood. It was dark, but I could make out an egg. I stretched out my arm, reaching for it, and then – it happened! A short distance away near the corner of the barn came that menacing sound of pure evil and sure death, "Meeeeeooooowwwwwwww!"

Luther dramatically imitates an evil-sounding cat, making sure that Joan and I are both giggling before he continues talking.

He says, "It was the Catman of Paris coming to get me! All at once, the hair on my scalp and neck tingled and stood straight up. I flung my bucket of eggs. I ran back the way I came as fast as my skinny legs would move. I burst through every spider web in the loft, stumbling and tumbling through the loosely stacked bales. Blindly, through the dark, I fell out of the loft. A pile of loose hay broke my fall. In an instant, I jumped up and ran straight toward the back door of our house."

Joan and I are laughing; Luther seems to be afraid of the Catman even as an adult. He has dealt with thousands of wild cattle and a few barroom brawls, but apparently he has a weak spot for a cartoonish cat-like creature.

He declares, "To this day, I will not watch a scary movie and I do not like cats!"

"Oh dear," I say. "You are so dramatic! I think this is enough storytelling for one night. I've got to get going now."

"Alright," Luther says, "but, watch out for the Catman on the way to your car!! Meeeeooowwwww!!!!!"

CHAPTER 10

HORSE TEAMS TO TRACTORS

I decide to visit my grandparents again and I call ahead to make sure Luther is in the mood to tell stories. Joan answers the phone, so she hollers at Luther from the kitchen. He is watching TV. It is loud enough to hear over the phone.

"Luth," she says, "Sierra wants to come over. You want to tell stories?"

After repeating herself and waiting for him to "turn down the damn TV," the three of us finally decide to have supper.

Joan says, "I've got spaghetti, see you in a bit."

Everyone loves Joan's spaghetti. Her recipe is unique because she uses her home canned tomatoes, green peppers, onions and a large cast iron skillet. I can't wait to have a bowl.

Over supper, we talk about school, sports and my friends. Luther likes to keep close tabs on my girlfriends. He knows where they are attending college, what their majors are – and most importantly – who their boyfriends are right now. For a grandpa, I think he is pretty nosey. In a weak attempt to deflect Luther's attention from a friend's recent breakup, I ask him, "So, what stories will I hear tonight?"

In a booming voice he says, "Well, I decided to tell you about...*The Animals We Have Known*."

Kenneth Henry

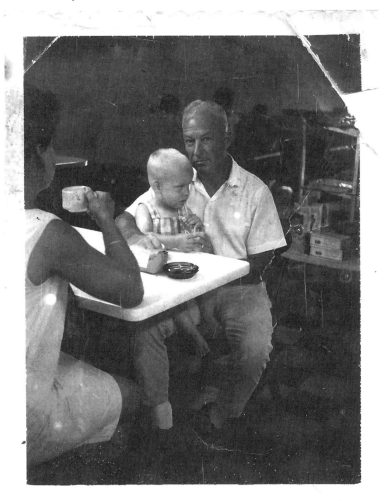

He says this phrase, '*The Animals We Have Known*', as if it could be a blockbuster movie title glittering in marquee lights. I have no idea what to expect. Hadn't he already told me about all the barnyard animals at his childhood home?

"You see, when I was a kid," he says, "horses were just going out of fashion. All of our neighbors were buying tractors for their fieldwork and farming. Since people didn't need to own as many horses, the price on a horse had really gone down. Within ten years of the introduction of tractors, horses were pretty much of no value to the farmers. Before tractors, a person could sell an old horse at the livestock auction at the end of its useful life. It might bring a good price. After tractors started to replace horses, the price of an old horse just wasn't worth the trip to the sale barn. People started to keep their old horses, waiting for them to die out on the farm. L.W. spread the word that we would come out to the farm and skin the hides off them dead horses. He told folks to call us when an old horse died. We would come out to the farm, skin the horse, sell the hide and feed the carcasses to our hogs."

"Oh dear," I think to myself as I take notes. I may not enjoy this set of tales. I have a soft heart for horses. This isn't the first family story I've heard about skinning dead horses. Joan's mother, Christine, married Kenneth Henry. The Henry family raced greyhound dogs; some still do in 2014. Occasionally, "Papa Kenny" was given a dead horse carcass that he could use to feed the dogs. Once when Luther was dating Joan while she was still in high school, he stopped by her house to see her. It was a surprise visit; he didn't call before he pulled up to her house. He found her in the backyard in a pair of shorty-shorts skinning a dead horse to feed the dogs. Skinning and slicing horse meat - what a date!

Luther doesn't mention if he thought skinning dead horses was a tough or unpleasant job. He knows this is a story that dates him. These are stories about animals he knew, horses made their exodus as part of the standard farm landscape when Luther was ten years old in about 1944. While textbooks may record the transition from plow horses and mules to tractors, they don't include much on skinnin' horses. It was probably too unpleasant. Prior to this time, horses provided a work force to the agricultural community. In this area today, horses are pets.

In this era, there was an unwillingness to waste on the farms and in rural America. Luther and Joan also display indifference towards dirty, physical, unpleasant work. They simply did the work that needed to be done.

A pair of greyhound dogs owned and raced by the Henry family.

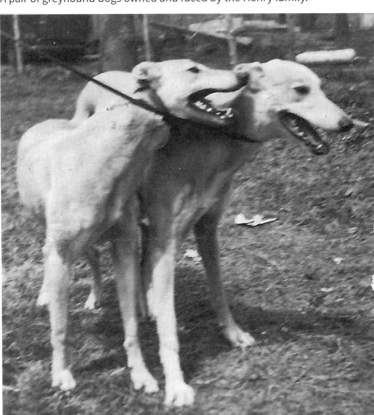

CHAPTER 11

SLOP FOR THE HOGS

"During World War II," Luther continues, his face gently lit by the reading lamp beside his chair, "L.W. came up with another big idea for us kids. He decided that our family would collect all of the edible garbage from the restaurants in Centralia. Then we would feed the scrap food to our hogs."

"Free feed!" I chime in, remembering how L.W. enjoyed lowering his cost of production. It was a survival skill he learned after his father passed away. In time, it simply became a habit.

Luther smiles and says, "He put a fifty-gallon barrel in the alleys behind each restaurant's kitchen. Then, he worked it out with the cooks and the busboys. They would throw the scraps into those barrels. To collect the scraps, we used our family car. Not a feed truck, our family car!"

L.W. thought this plan would work out well. With it being

wartime, no one could justify wasting anything – even food scraps. Many of the busboys were happy to take the extra steps out to the back alley to toss the scraps. It was damn near patriotic. Louise didn't have much to say when they started using the car to haul the scraps; it was just easier to get around town that way.

Luther continues, "L.W. would take the seats out of the back of the car to make room for all of the five-gallon buckets we filled by dipping the scraps out of those big barrels. We drove down the alley and pulled the lids off the barrels. They were full of kitchen scraps that were two or three days old. Collecting and feeding these scraps was a bad job in the wintertime because of the stink, but in the summertime it was a really bad job! The summer heat soured the scraps."

As Luther stared over the edge of barrel, he cringed at the sight below. There were half-eaten hamburgers, soggy fries, pancakes, cottage cheese and crusty dinner rolls – and that was just the top layer. After L.W. pulled another lid off, an odor – a mixture of a week's worth of diner food – escaped. Luther resisted the urge to gag.

Luther says dramatically, "You just can't imagine, much less describe, that kinda stink. What's worse, you can't imagine the mass of flies that them scraps attracted. You just couldn't get away from that many flies. We spent our mornings driving all over town collecting scraps and then we drove out to the hogs. At the time, we were keeping our hogs a mile or so west of town. L.W. had a twelve-inch plank that we laid across the buckets. He instructed us kids to sit on the board in the back of the car. Our weight and the board kept the slop buckets from tippin' over while he drove. During all that driving around town, it seemed that every day, some of that soured slop would splash over the edge anyway. Of course, it sank into the carpet of our family car. After a while, the back of our car got to be pretty rank. We always drove with the windows down, leaning out. I felt a little bit like a dog, trying to get some fresh air. I bet there was a green stink cloud that followed our car, just like a cartoon. If it wasn't stink...it was the damn flies following us from town in a black cloud!"

Luther is getting really revved up now. If he were telling this story in a barroom, the crowd would be zeroed in on him and laughing at the thought of ten thousand flies following the family car.

Luther continues, "When we got to the farm, we ran the hogs out of the lot. We had to get them out, so we could pour all the slop into several large wooden troughs. If the hogs were in the pen, they would knock us over before we got the slop spread out. To make this job worse, somebody went and told L.W. that hogs would choke and die if they ate fish bones. After that, he started making us sort through the slop. We had to run our hands through the scraps after we dumped them into the wooden troughs."

Louise likely had to hose her dirty children off in the backyard before allowing them to come into the house to clean up and bathe after this dirty job. The cold water from a spouting hose might have sprayed away the leftover food scraps that were stuck to their clothes, making Louise's washing easier, too.

He continues, "The flies would still be giving us fits. It didn't take long for hundreds of them to swarm around me. I'd sort and pick out the bones as fast as I could, but the flies would get to buzzing in my ears. Those flies were enough to make me crazy!"

He waves his arms around his ears and face wildly, as if a swarm of flies has just invaded the peaceful living room.

"They were so thick and aggressive, they would go in my nose and I'd be spitting them out my mouth," he explains. "There was nothing I could do when them flies landed on me, tickling my skin, because my hands were covered in slop! So, we just tried to be tough while a cloud of flies tickled our faces, pickin' fish bones out of that sour, stinky slop. L.W. was sure happy to take advantage of that opportunity, since we was feeding our hogs for free. But for me, I still remember it as a pretty bad job. Looking back, I don't know why he didn't just ask the busboys not to put the damn fish bones in the slop barrels in the first place!"

CHAPTER 12

PICKIN' WOOL

Without mentioning it, Luther's boyhood stories take on a theme. His childhood was dominated by the work L.W. created for his children. The work he found was always stinky, causing most folks to avoid it simply because of the unpleasantness. L.W. was hard of hearing, but did he also lack a normal sense of smell?

Luther says, "Back in the '40s, there were still a lot of sheep in Missouri, too. L.W. did a lot of sheep selling and trading with the neighbors. The sheep came up from Texas into Centralia on train cars. The train depot in town had a small yard, but most of the time, L.W. would just back one of our straight trucks right up to the railcar and Buddy and I would run the sheep right onto the truck."

There is truth in Luther's fancy-sounding title: *The Animals We Have Known*. In his lifetime, he and his neighbors saw the exodus of three major species of livestock from nearly every farm in Missouri – horses, hogs and sheep.

The differences don't just stop at livestock species. The methods for hauling were far different, too. Thanks to the modern interstate system, livestock haven't been hauled in or out of central Missouri by rail in decades. A story about sheep hauled into Centralia by rail hardly seems possible.

Luther continues, "Then, we sold and delivered the sheep to people around town. Somewhere in my house, there is an old boot box full of receipts from when L.W. bought sheep out of Texas."

"Joan," Luther bellows, "where did that box of sheep records go? Are they in the attic? I know they are in an old Nocona boot box."

She is sitting just a few feet away, but he still yells to hear himself speak. She says calmly, "You've been asking me about that box for years. I always tell you it is somewhere up there in that attic."

It flusters her when she can't find things, but over the years, these two have developed quite a collection of items. Joan, a woman with four or five crock-pots, likes to remind us, "The only thing I have one of is Luther!"

Even though we don't have the old records, Luther can still tell the tales to me and Joan. He starts again, "During the late summer, people would turn feeder lambs out into cornfields. There weren't many chemicals used for weed control back then. The sheep would eat down the weeds instead. If you pulled the sheep out of the field soon enough, they wouldn't even eat the corn off the stalks. But if you wanted them to get real fat, you could leave them in the field longer to eat some of the corn."

It amazes me that farmers used to allow sheep their cornfields. Since the 1990s, most area farmers have torn out the fences around their fields.

"With all them sheep around, a few always got killed by the dogs and coyotes," he continues. "During the war, you could sell just about anything. Everything the Army needed was in short supply. Me and my brother and our friend, Roy Ellis "Shorty" Hulen, worked together a lot. We figured out that wool was worth something like a dollar a pound. The Army was still using wool to make the topcoats and pants for soldiers' uniforms. That was a high price for wool. It might be even more than it is worth today. So, we started picking the wool off of our sheep that died and selling it."

L.W. didn't come up with all of the work for his kids. He also taught them how to find opportunities for themselves. After skinning a dead horse and collecting soured food scraps for hogs, pickin' wool off a dead sheep was just a regular job to Luther and Buddy. After L.W.'s training, this was another prime opportunity to make money with a stinky disguise.

Luther says, "Pretty soon, the neighbors figured out that we were wanting the wool off them dead sheep. They started calling us when they had one or two die. Then we would come out and pull the wool off for them. We'd sell the wool and split the money."

Luther leans back into his chair and smiles. I sense he is working up to a punch line.

He says, "Now, pickin' wool ain't easy! Not just anybody can do it. There is an art to pickin' wool off a dead sheep. I wish there was a dead sheep here right now so I could show you just how to do it." He points right at me, letting me know this is an art I should learn.

I laugh. "I'm glad there's not a dead sheep in this living room!"

Luther continues giving instructions. He speaks as if tomorrow, I may need to know how to complete this process. We both know that it is very unlikely I'll need to pull wool off a dead sheep, but it makes for a good story.

He says, "You gotta know when the time is just right. If you try too soon after they die, the wool won't come off good. If you wait too long, their bellies swell up full of gas. Then, you go to pull the wool off—pop! Them sheep bust open."

77

He ducks and covers his head, yelling the word, "Pop!" If one of the boys pulled wool off a sheep that was too ripe, the explosion would send guts, half digested food and other slimy substances flying. When they heard the noise, all three hit the ground, hoping not to get hit. Then they'd slowly rise up again, checking their clothing and hair to see where the guts had landed. They checked each other over, too, and then went back to work.

He continues, "Now remember, you gotta learn when to pick 'em when they are just right. If they are ripe, the wool would come off nice and easy – no popping open!"

Luther laughs and his round belly shakes. He has gotten many a crowd to laugh at that story, in country clubs, sale barns, barrooms and restaurants. The higher class the crowd, the more fun he has telling the country bumpkin-style tale. Once my laughter calms, Luther mentions quietly, "We made lots of money pickin' wool off of dead sheep."

Suddenly, L.W. turns from a relentless man who provided a childhood of tough work for his kids into a wise father. L.W. gave his children their independence by teaching them that no work was too tough or too low. L.W. taught his kids how to make money.

Luther adds a bit more to his story of pickin' wool, an oral epilogue of sorts. "Now, my friend Shorty was the youngest of five boys. He always helped me and Buddy pick wool. He was the only one in his family to get a high school education. Over the years, Shorty worked in several of the banks in Centralia. He started as a clerk and moved his way up until he eventually ran the bank. He never went to college, but his life experience made him a real good banker. I told that sheep pickin' story on the radio a few years ago during a remote broadcast at the western store. Some lady was listening and recognized "Shorty" as her banker's name. She just couldn't believe that story was true about the man overseeing her money. So she called him up at the bank office. 'Shorty,' she asked, 'did you really pick dead sheep when you were a kid?'"

Luther is excited to tell a tale on his good friend Shorty Hulen.

"Now," Luther says, "some bankers may have been inclined to cover their tracks or downplay such a colorful event when confronted years later by an influential customer. But not Shorty! 'Yes!' Shorty told her proudly. 'I did pick wool with ol' Luther....'"

Luther muses with a twinkle in his eye, "If we had a few less college-educated bankers and a few more bankers who had been educated on more humble pursuits like pickin' dead sheep, I bet we would have fewer banking problems in this country!"

1948-1952
HIGH SCHOOL

This Deed, Made and entered into this ___26th___ day of ___March___

A. D. One Thousand Nine Hundred and Forty -Eight___, by and between ___Luther W. Angell and___

___Mary L. Angell, his wife; Carl S. Hulen and Catherine B. Hulen,___

___his wife; and Frank Elkins and Nena P. Elkins, his wife,___

of ___Boone___ County, State of ___Missouri,___ part__ies__ of the first part, and

___Columbia Livestock Auction Company, a corporation,___

of ___Boone___ County, State of ___Missouri,___ part__y__ of the second part:

WITNESSETH, That the said part__ies__ of the First Part, for and in consideration of _____

TEN AND NO/100 - DOLLARS

___and other good and valuable consideration___

to ___them___ paid by the said part__y__ of the Second Part, the receipt of which is hereby acknowledged, do_____ by these

presents Grant, Bargain and Sell, Convey and Confirm, unto the said part__y___ of the Second Part, the following described

tract___ or parcel___ of land, situated in the County of Boone, in the State of Missouri to-wit:

A tract of land in M. R. Conley and Wellington Gordon's
Subdivision described as follows: Beginning at an iron
set on the West line of Lot 14 in said subdivision and
on the South right-of-way line of U. S. Highway No. 40;
thence with the West line of said Lot 14 South 561.3
feet to a stone; thence North 76°-31' West 298.2 feet to
an iron; thence North 492 feet to an iron on the South
right-of-way line of U. S. Highway No. 40; thence with
said right-of-way line South 89°-57' East 290 feet to
the place of beginning, being a part of Lot 13 and the
closed street adjoining in M. R. Conley and Wellington
Gordon's Subdivision in the Northeast Quarter (¼) of
Section 7, Township 48 North, Range 12 West.

(Actual consideration less than $100.00).

TO HAVE AND TO HOLD the same, together with all the rights, immunities, privileges and appurtenances to the same be-

longing unto the said part__y___ of the Second Part, and to ___its successors___ ~~heirs~~ and assigns, forever; the said

___parties of the first party___ hereby covenanting that ___they and their___ heirs,

executors, and administrators, shall and will WARRANT AND DEFEND the title to the premises unto the said part__y___

___successors___

of the Second Part, and to __its___ ~~heirs~~ and assigns, forever, against the lawful claims of all persons whomsoever

___except all taxes falling due for the year 1948 and thereafter___

IN WITNESS WHEREOF, the said part__ies__ of the First Part ha__ve___ hereunto set ___their___ hand_s_ and

seal_s_ the day and year first above written.

WITNESS

[signatures]
Luther W. Angell (Seal)
Mary L. Angell (Seal)
Frank Elkins (Seal)
Nena P. Elkins (Seal)
Carl S. Hulen (Seal)

The original deed for the Columbia Livestock Auction property.

CHAPTER 13

THE COLUMBIA
LIVESTOCK AUCTION

"When did L.W. get started in Columbia?" I ask Luther. We are sitting at Angell's Western Wear. This Saturday afternoon, it is rainy and business is a little slow. The western clothing store was another one of L.W.'s original business ventures.

Luther says, "In 1948, L.W. and a few partners bought the Columbia Livestock Auction. The barn and the land cost $20,000." When Luther starts talking about the Columbia Livestock Auction, he beams. His happiest childhood memories, and a few of mine, began at the Columbia Livestock Auction.

"They had four partners, so each man had to come up with $5,000," Luther says. "That was a lot of money. It was hard to come up with that much. The day they were supposed to close on the property, one of the partners backed out. The remaining three partners had to scramble to scrape up enough money for a three-way split. The three partners were L.W. Angell, Carl "Doc" Hulen and Frank Elkin. It was stressful, but they got the barn bought."

Luther was fourteen years old when L.W. purchased the Columbia barn. Luther may forget names and dates, but he does have great skill in remembering prices. It is no surprise that he remembers the cost of the facility.

Luther says, "After they bought the barn, me, Buddy and Rosemary spent the rest of the summer helping get ready for the first sale. We painted and cleaned and picked up junk. The first sale was held on a Wednesday. It was a small, community sale barn. The area around Columbia had a lot of dairies, so we sold a lot of dairy calves."

The late 1940s were an interesting time for enterprising farmers and cattlemen. The country was recovering economically from the Great Depression and, more importantly, people were recovering emotionally. They were ready to take risks again. It seems that L.W. was ready for risk in 1948. That year, he purchased a one-third interest in the Columbia Livestock Auction. He also purchased three grass farms around Centralia. Unlike previous deals with L.C. Henderson, L.W. held onto these farms the rest of his life. He purchased them from the Metropolitan Insurance Company, meaning they had been lost by area farmers during the Depression.

After he passed away, the farms were divided between the three children. Luther lives on the home place next to the golf course, Buddy lives on one south of Centralia and Rosemary rents her farm out.

The Columbia Livestock Auction in 1947.

Luther was so thin and gangly that his friends nicknamed him "Frog Legs."

Back row, left to right: Assistant coach, Ed Kenne; 57, Junior Palmer; 59, Darrell Chase; 51, Red Jennings; 53, Luther Angell; 55, Johnny Carter; head coach, Pete Adkins.

Front row, left to right: 52, Don Griffin; 50, Howard Lewis; 56, Jerry Cox; 58, Mike Way; 54, Jimmy Northcutt

Coach Pete Adkins started his legendary high school coaching career in Centralia. In his career, mostly with the Jefferson City Jays football program, he won 405 games, with a 71-game winning streak. He also captured nine state football championships and a spot in the Missouri Sports Hall of Fame.

CHAPTER 14

THE MOONGLOW

Luther says, "One time, two of my upperclassmen friends, Charlie and J.R., took me out driving. I felt pretty cool hanging out with the upperclassmen; those older boys took me under their wing. They sorta showed me the ropes of high school. One day, the three of us left Centralia heading east toward Thompson."

"Luther," they said, "we're gonna teach you a magic trick."

The Moonglow was a roadhouse bar and dance hall east of Centralia near Thompson. It sat alone on Highway 22. The owners were Charlie and Grace Fry, and there were no other businesses around.

"When we made it there, one of the older boys jumped out of the car and run up to the building. My friend knocked on the back door, set $1.25 on the top step and then ran back to the car. He got back in and we drove off down the gravel road. A few miles and minutes later, we turned around and headed back toward The Moonglow.

"While we were gone, the money turned into a six-pack of cold beer. It was waiting for us right there on the top step. From then on, I could get a cold beer just about anytime I wanted. Thanks to my friends, I knew how to do the magic trick, too."

In the 1940s and 1950s, The Moonglow sat on the corner of Highway 22 and a gravel road. Today, the gravel road is a paved state blacktop road officially named State Road E. The locals, even the younger ones like me who never even saw the bar before it was torn down, still call it "Moonglow Road" instead of "Highway E".

Luther says, "When I got brave enough, I stopped knocking on the door and leaving money and just started going inside with my friends for a few drinks. We danced with all the girls and played our favorite songs on the jukebox. I had a real good time out at The Moonglow; I could get a pitcher of beer for $1.25. They also served a plate of crackers and cheese for fifty cents."

"It seems pretty cheap!" I say. "You can hardly buy one beer for that price today, much less a whole pitcher."

"Yep," Luther says. "One night, me and the regulars were out at The Moonglow when our friend Tommy walked in wearing a brand new sport coat. We all agreed he looked pretty sharp. Even Charlie and Grace thought Tommy's coat was pretty fancy. After we gushed about Tommy's coat, the night went on like usual. That meant...we all drank too much. Tommy's folks were not home, so we decided to stay there after the bar closed. We headed towards his house with several of us stuffed into two or three cars, bouncing down the gravel roads.

"Tommy's drinks and that bumpy road did not mix well. He turned green. Before we made it to the shoulder of the road, Tommy threw up all over himself and the car. His new coat was a mess. There was nothing to do about the sorry situation now, except to get Tommy home in a hurry. We arrived and I helped haul him into the house. Tommy's stinky coat needed a little air, so one fellow hung it in a tree out in the front yard. We got Tommy cleaned up and then everyone fell asleep.

"The next morning when we woke up, I looked out the front window and saw big black birds all over the yard. Tommy's coat was hanging on a branch and all them black birds were circling around it. Those birds pecked hundreds of little holes in Tommy's new coat! They were eating all the puke right off his jacket. After that, Tommy learned to wear his older clothes when he was out at The Moonglow."

Luther says, "The funny thing about The Moonglow was that all the high school kids from Centralia and Mexico went out to that bar. Since it sat right between our two towns, both schools went out there. At the time, Centralia and Mexico had a big sports rivalry. It was not unheard of for a fight to break out during a football game or afterwards out in the high school parking lot.

"Since kids from the two schools did not get along, they sat on one side and we sat on the other. People from the two schools never talked or mixed much at that bar. Surprisingly, there were never any fights out at The Moonglow between kids from rival towns."

I suggest, "I think you all knew you had a good thing going out there; nobody wanted to mess it up with a fight."

Luther says, "Of course, it was different times back then. When I was in high school, we was fighting in the Korean War. The laws said a person had to be twenty-one to buy and drink alcohol, but no one followed that law. The unspoken rule around was that once you was tall enough to put your money on the bar, you got served a drink. It was not supposed to be that way, but it was. Eventually, while in high school, I met the owners, Charlie and Grace Fry. They were nice folks. We became friends and every year they sent me a Christmas card. This was funny because my mother always said, 'Luther, who are the Frys? I don't know them and they sent us a card.'"

Like a typical teenage boy, he just shrugged his shoulders and shook his head.

"I don't know, Mama," Luther said, "I don't recognize that name either."

CHAPTER 15

CARS AND OIL

"You know, Sierra, when I was your age, girls didn't get to have cars?" Luther tries to irritate me, "They had to ask a boy for a ride! What do you think of that?"

Joan and I roll our eyes, ignoring his teasing.

"Sounds pretty silly to me," I tell him. "I can drive just fine."

I'm grateful that I grew up in an era where sixteen-year-old girls were allowed to own cars. I didn't have to wait around for a boy in order to get someplace.

"I got a car though!" Luther grins, knowing he has ruffled my feathers. "It was a 1932 Chevrolet. It only cost sixty dollars. Actually, the car didn't even cost me that much. Me, Eddy and Junior each put in twenty bucks. The three of us shared the car."

Today, folks pay more for a tank of gas than Luther did

for his first car. I envy those old models because they were impossible to hurt. The steel bumpers were actually made for "bumping," unlike modern plastic bumpers that crack if a shopping cart accidently rolls into the back end.

Luther says, "In high school, the cool thing to do with an old car was give it a wild paint job. We painted our car with pokey-dots, stripes and crazy patterns. It had a different color on every panel. All the kids around town painted up their cars; it wasn't just us. The high school parking lot looked like a rainbow of paint cans had exploded on the cars! Why don't you do that to your car?"

"Well," I say honestly, "I guess it never really crossed my mind."

"You kids are no fun," he declares. "I think you all ought to paint up your cars. It is fun. After I got my license, one of the highlights of my high school summers was the Green Girls coming to visit. The Green Girls were from New Madrid, that is down in the boot heel of Missouri. Every year their parents sent them up to Centralia to stay with their grandma and their great aunt for most of the summer. After a whole school year looking at all the same girls, the guys and me were pretty excited when Mary Lee and JoAnn arrived in town. The Green Girls were cute and fun, so they got plenty of attention from all the boys in Centralia. Their arrival did not go over well with the hometown girls. They were all jealous of the Green Girls because they stole the boys' attention every summer.

"On the weekend, my buddies and me would take Mary Lee and JoAnn Green out driving with us in our painted-up car. Their grandma and auntie made it a rule that they had to be home by eight o' clock. At first, we always got them home on time. Then, one night, we came up with a more appealing plan. We dropped the girls off, drove around town for a half hour or so and then parked behind their house with the lights turned off. While we waited, the girls told the old ladies good night and then pretended to get ready for bed.

After awhile, they shut off their bedroom light and snuck out of their window. They ran quietly across the lawn in the dark. Once they made it to our car, we drove off again. By 8:45, us boys and the Green Girls were out having fun again!"

Luther wants to drive right back into those teen years, where he can hoot, holler and speed away after successfully sneaking the girls away from the house in his sixty-dollar car.

He says, "We spent lots of time driving around and having fun together. Those old cars had one flaw; they took a lot of oil. I kinda lucked out on that end of the deal though. My friends Eddy and Junior both had jobs working at the filling station. It was a full-service station, so when a car pulled up to the pump it was their job to fill it with fuel. They also cleaned the windows, shined the tires and made change for the driver. They didn't get tipped for doing these extra jobs; this was just the normal routine at a full-service gas station. The driver never even got out of his car. Sometimes, they would be asked to check and change the oil. When they changed the oil for a customer, they always drained out the last little bit into a different pail. The rest may have been gunky, used-up and full of dirt, but that settled at the bottom of the oil pan. But the last bit was usually pretty good. Rather than throw it away, my buddies saved it for the car we shared. We needed so much oil to run that old car, we kept a five-gallon can of oil in the trunk all the time. I bet it took us one gallon of gas and two gallons of oil just to drive the twenty miles over to the next town!"

CHAPTER 16

THE CENTRALIA HOG BARN HISTORY

Luther yells down the alley to me, "In the Holstein! By the black!"

He stands near the group of about thirty cattle and allows one or two at a time to come down the narrow alley toward me. If he shouts, "In!" I let the calf into the pen. If he shouts, "By!" I close my gate and let the steer pass me and go further down the alley.

He yells, "In the black one! Watch him! He's coming fast!"

I hustle to open my gate. If I'm too slow, the steer will end up in the wrong group. Or worse, the steer will run right over the top of me! Luther wants this steer to go in the pen, so my gate needs to be open. We are sorting Luther's cattle at the hog barn. The facilities in the back are used frequently for this purpose. The small, dirty office isn't the only section of the building that serves a good purpose. Luther buys cattle that weigh four to six hundred pounds. Then the cattle are grazed until they weigh about seven to eight hundred pounds. When the feedlot out in Kansas has an open pen and Luther thinks the time is right, we gather the cattle from the pasture. We haul them to the hog barn to sort, weigh, load and ship them.

Shipping day is fun. On Saturdays, the grandkids – including me – can help. It is an "all hands on deck" kind of job. Since the Angell family doesn't have cowherds, we don't do any branding. The practice of branding is not common in Missouri either. However, I think a morning of shipping cattle is as close to a branding atmosphere as we get. It is part work, part social event. It doesn't matter if they are your cattle or not – you show up and help. In this way, everyone in the family swaps labor. Next time your feeder cattle are ready to ship, you'll have plenty of help, too. A couple of trucks and gooseneck trailers are scheduled to haul Luther's cattle into town at seven or eight o' clock. One or two people are stationed at the farm to help each driver load the cattle. Next, the cattle are hauled to town to the hog barn and unloaded, sorted and weighed.

According to Missouri's state laws, a semi-truck and trailer loaded with livestock can only weigh 80,000 pounds. After subtracting the weight of the truck and trailer, we get to load about 50,000 pounds of cattle.

After Luther has sent ten to fifteen head "by," I run the cattle down the alley to the scale pen to weigh them. I open the wooden door into the scale house and rush in to see what they weigh. The scale is more than sixty years old now, but it still works well. The scale house is made of white wooden boards. They are covered in scribbled math problems from decades of doing math to properly load trucks.

This scale isn't electric, so the weight doesn't appear on a screen instantly. There is no screen. Instead, the four-foot metal contraption functions with a weight and counter weight on each end. A long numbered arm shows the weights by hundreds – 100, 200, 300, 400, 500. When the long numbered arm balances evenly, I've found the right weight.

I hastily record the weight of the draft and the number in the group in a notebook. Unlike some of the others who have used this scale house, I have a notebook so I can write my figures on paper, rather than on the walls.

If things go well, Luther's loads will work out almost perfectly at 50,000 pounds. After we weigh several drafts, I add up the figures to determine how close we are to 50,000. If the number is 48,000, Luther will sort out a few more to add to the load. However, we don't want to be overweight because the truck driver could get a ticket. When the load is at 50,000 pounds, Luther starts doing the math in his head to load the trucks. First, about 5,000 to 6,000 pounds can be loaded in "the nose," which is a small section at the front of the truck. Then, about 20,000 pounds can be loaded in the top and 20,000 in "the belly." Finally, at the back end of the truck, another 4,000 to 5,000 pounds can be loaded.

He barks, "They are averaging 750. How many can we get in the nose?"

This is a test of my mental math skills; I'm awful when put on the spot. I try to hold the invisible numbers up in my mind's eye, dividing 750 into 5,000. But, soon I've forgotten what number was in the tens place.

Luther says more gently, "If they weighed 700, then ten would be too many. So, we know it is less than ten."

He is trying to be encouraging. It's not helping; I'm still awful at mental division. Standing in an alley with cattle bellowing and people waiting doesn't help.

"It is seven up front and six in the back." Luther continues, "Okay, now, we've got fifty-three head left. Now how many go on the top and how many in the belly?"

I'm relieved. I can add and subtract in my head easily.

I think to myself, "Sixty-six minus thirteen equals fifty-three and fifty-three divided by two is…."

"There should be twenty-six on top and twenty-seven below," I state proudly.

With that job finished, he turns to see if the truck driver is ready for us to load the cattle. I'm trying to remember: 7-26-27-6. We need to get the truck loaded fast because there's another truck waiting. However, even if there were only one truck, we'd still be working fast. Because, well, that is what we always do!

We load two trucks with 50,000 pounds of cattle each. Then, Luther has to write up a "bill of lading," which includes the ownership, headcount and weight of the cattle. Although we loaded 50,000 pounds of cattle on each truck, he uses the industry standard two percent "pencil shrink." This helps account for the weight the cattle will naturally lose during shipping, giving the feedlot a more realistic beginning weight of 49,000 pounds. This will provide a more accurate view of the cattle's performance in the feed yard.

He also includes the health papers, which are required for the cattle to travel across the state line from Missouri into Kansas. Then he'll give the paperwork to the truck drivers and they will be on their way.

We meet in the office at the front of the building and Luther gives my cousin, Jensyn, and me forty dollars for breakfast. It is way too much money, but he's always generous. We take orders from everyone who helped and then head into town to the Sonic drive-thru. We come back with large Cokes, iced tea and a half dozen breakfast burritos. The trucks are gone when we get back; they have a long drive. Buddy has also stopped by the hog barn after finishing his morning rounds checking his grass cattle.

It is only 10:45 a.m. Lots of our friends from school are still sleeping, but Jensyn and I feel like we've gotten in a half a day's work already. I pass out drinks and Jensyn tosses bacon or sausage burritos out to hungry helpers. Like a branding, a meal is always involved with shipping cattle. Whoever owns the cattle buys breakfast. Everyone lounges

on couches and chairs. The floor of the office becomes dirty
with clumps of manure from our boots. The couches and
chairs are full. The small room feels crowded, as if we're
having a party.

Since Jon, Justin and Luther started early shipping cattle,
they have not had time to do their chores yet. They finish
breakfast and make plans for the day.

Luther says, "Jensyn, you got time to help me get a steer
up to doctor?"

Jensyn nods his head yes and asks if they need to pick up
any medicine. Luther and Jensyn pull out of the hog barn in
Ol' Red to go and check his cattle. They head east toward
Luther's house, probably to pick up the medicine. Ol' Red
has been Luther's feeding pickup for the last decade or so.
There are enough dents, dings and scratches to make the
shell of the truck resemble the dimpled hull of a golf ball
more than a smooth hood.

Justin pulls out of the driveway behind his father. His
pickup is in just slightly better condition. No one can wreck
a truck quite like Luther. Justin's pickup is red with a black
flat bed and bale spears. It has plenty of dents and one of the
door handles can only be opened from the outside. To get out,
the passenger has to roll down the window and reach to the
outer handle. Jon jumps into a little blue Geo. When RTVs
came into popularity, Jon realized that rather than buying
a ten or fifteen thousand-dollar RTV, he could find a decent
used Geo for a fraction of the price. The Geo is a smaller,
off-brand version of a Jeep. When he is not hauling bales, the
little vehicle has proved quite handy.

Buddy and I had planned to talk more, so we stay behind.

"So," I ask, "how did the Angells end up with this
building?"

Buddy says, "Well, twenty-five farmers from around the
area decided to pool their money and build a livestock auction for the town of Centralia. They devised a system where
everyone could invest different amounts of money. They

started fundraising in 1949 and broke ground in 1950. Luther was a senior in high school at the time. His agriculture class left school for a field trip. They attended the very first sale."

The community was very proud to have their own live-stock auction. The auction office where we always sit to visit has one corner sectioned off with counters and windows for buyers and sellers. The office staff would have done all of the paperwork within the bars. The small windows were for passing cash, checks and buyers' numbers.

Buddy says, "The sale didn't last more than five years though, because all the owners wanted to have a preferential job. They were all fighting to be the ring man, clerk or office manager. No one wanted to work out in the back bringing up cattle, loading, sorting and penning. The large group bick-ered all the time; there were too many chiefs and not enough Indians. Within five years, they decided to shut down."

It was disheartening to see their new building closed down. The investors were stuck with their money in a brand new building that was now closed.

Buddy said, "Although the sale didn't work out too well, that group did one thing really well. For about five years, they held an annual sheep sale every fall. The whole commu-nity came together to help hold the huge sale."

Buddy realizes that I have never done any work with sheep. There are probably less than one hundred sheep in Boone County in 2014.

He explains, "You see, in the '50s, it was not a real big investment to buy a feeder lamb—about twenty dollars a head. When the lambs grew and got fat, they sold for about forty. Folks made a good profit on lambs, because they ate the feed left out in the cornfields that the pickers missed. E.B. Wilson organized most of the sale because he owned about seventy-five percent of the lambs that would sell. His brother-in-law sent hundreds of rail cars full of sheep up from Texas to Centralia. A double-deck car held two hundred feeder lambs, a hundred head on each deck. The Centralia

Livestock Auction barn could never hold five thousand head of lambs. That's why it took so much help from the community to host the sale. We built temporary fences to hold the sheep. The temporary pens covered the grassy area to the east of the barn and the west side of the barn. We tied snow fence to steel posts and then we put wooden gates between the rows of snow fence. It was quite a bit of work, but everyone pitched in and we got it done."

Buddy points out the east windows to the grassy area that my cousins now mow. I try to imagine thousands of little lambs standing out there; now there are just a couple trailers and an old dump truck parked in a tidy row.

Buddy says, "At least five thousand head of sheep were sold each fall at the sale. Some years, I bet the number was maybe even as high as six or seven thousand. Folks came to the special sale from all over Missouri, Iowa and Illinois. That was one thing the Centralia Livestock Auction did real well. After they closed the barn down, they leased it to a man out of Arkansas for a while. He used it as a hog buying station. Then it eventually came up for sale and L.W. bought the building. At that time, L.W. was already working at the Mexico Livestock Auction and the Columbia Livestock Auction. He reopened this facility as the Centralia Livestock Auction. However, he only did that for a few months. Pretty soon, he figured out that the three barns were too close together. So he shut the Centralia barn down again."

The community of Centralia had tried and failed twice to have a community livestock auction. The large, well-designed facility was truly only used for its intended purpose for less than a decade.

Buddy continues, "After L.W. closed the barn down, he used it a lot like we do today for sorting and loading out cattle, and as a drop-off location for his customers at the livestock auctions. It was me and Luther's job to clean out the pens in the back of the barn for L.W. Our friend Shorty helped with that job sometimes, too."

For some time, the building served a good purpose as a sorting and depot-type facility.

Buddy smiles, "This was actually my favorite place to practice my auctioneering. Because I could use the old microphone system to practice while no one was around to listen. This made Luther real mad. He was out there scooping manure by himself and he wanted me to help. Eventually, he'd get tired of my chanting. He'd come up to the auction block and say, 'That is it! You are done! Get out here and help me!' Then he would take the fuse out of the microphone system. He wouldn't give it back to me until we were done scooping manure!"

L.W.'s certificate for a share in the Centralia Livestock Auction was issued in 1949.

CHAPTER 17

GEORGIA KINFOLK

A few days later, I visit my grandma Joan around 10:30 a.m. She grew up in Georgia and I know very little of her life there. I'm excited to hear about her family. My father, Justin, has briefly mentioned fun childhood memories of visiting Joan's Georgia family during the summertime. I pour myself a cup of coffee and hers is already sitting on the small warming plate. I warm up a piece of leftover rum cake, deciding that it will be my brunch. The cake is delicious and simple; the secret is slowly drizzling a rum glaze over the freshly baked cake.

After getting our coffee and cake, we settle in at the kitchen table. This conversation is more intimate than talking with Papa; his stories are geared for a crowd and lots of laughter. Joan's Georgia childhood tales come from an era and part of the country far removed from my life. I was born in the Midwest in 1991; she was born in the Deep South in 1937.

She says, "My father's name was Thomas Olin Gassett, but everyone just called him T.O. When I was growing up, me and Daddy lived in a big old house just about a block off the main highway through Roberta. The house had a long hallway going right through the middle. A little old lady owned the house, and she lived on the north side and we lived on the south side. We had three rooms. When you came in the front door, we had a company room. Then, me and Daddy had a bedroom. He had a big bed and I had a little twin bed in the corner."

Her description of the house is helpful. She knows we'll never visit this place together. The twelve- or thirteen-hour drive south through Missouri, Illinois, Tennessee and finally to the middle of Georgia would be fun though – what a road trip. In the 2000s, Joan and Luther traveled to Georgia and the house was still standing, although it had been moved a block away from its original location along Highway 341.

She says, "My parents divorced when I was about two years old, and from that time on, I lived with Daddy in that apartment. He hired a nigga woman to take care of me and keep house. There were four women over the years, but Aunt Belle was my favorite. She started about the time I went to school. Every week she stayed with us, she cooked, cleaned and did all the laundry. She had a little twin bed in the third room, which was the kitchen."

I know enough about the Southern culture during that time to know that Joan is not being racist when she uses the N-word. She is technically being politically incorrect and the hairs on the back of my neck rise to hear her say it, but she never speaks ill of Aunt Belle. There is only love in her voice.

The type of relationship she describes has become just as outdated as the choice of word. An African American woman was hired on to raise my grandmother. It is part of our family history and an important part of Joan's life. I had considered

WILD MAN: A MOSTLY TRUE MEMOIR OF A MISSOURI CATTLEMAN

editing this scene out, but that wouldn't do justice to the history. It is also an important part of Aunt Belle's life. Leaving her out of this story because that's more comfortable wouldn't benefit anyone. Aunt Belle gave up time with her own daughter and grandchildren to care for Joan. I want to respect her sacrifice, too. She took the job to save money for her daughter's six children. Throughout the years, she helped send three of her grandchildren to Detroit with her savings. She believed that they would have a better life in the North.

Joan has no idea she's already given me plenty to think about; she simply explains the way things were. Before I've processed her relationship with Aunt Belle, she is moving on to describe more of the rented house.

Joan says, "The kitchen had a tall wooden cupboard, a wood cookstove, a table and Aunt Belle's twin bed. Every morning she'd get up and make her bed and then start the day. Nothing delighted me more than to run and jump in the middle of her feather bed. I wasn't supposed to, but sometimes I did. She wouldn't have a wrinkle in it and then I'd jump right in the center."

She says this as a confession, wishing she could go back in time and be less naughty as a young girl. She continues, "On Saturday mornings, Dad would take Aunt Belle out to her daughter's house. She'd stay with her daughter all weekend. Then, on Sunday nights he went back out to get her again and she'd come to work all week. Aunt Belle raised me. I thought of her just like a mother."

Joan takes a moment to sip on her coffee. Her diamond rings and bracelets sit in a small china dish on the kitchen table. Near that, a Lazy Susan contains paper towels in place of napkins. A small green bowl holds softened butter. The bowl is covered with a mismatched pewter lid. The small handle on the vintage covering is a statue of a cow. Joan liked the unique lid. An antique-loving neighbor searched for a matching bowl at antique shops for nearly a year. Joan's

friend finally gave up and hired an artist to make a custom-sized, dark green clay pot. Salt, pepper, vinegar with hot peppers, Louisiana hot sauce and a small airtight container for Luther's hearing aids fill the edges of the round turntable. A glass bottle with a slender neck holds tiny bright red cinnamon candies.

"When I was about two years old," she continues, "Daddy drove a gas truck all over town and out in the country. A few years later, he became a foreman at a sawmill. Every morning, he would get into his straight truck and drive away. He picked up many of the workers on his way. At night, he would drop everyone off again. It seems like I remember him getting home awful late in those years. They would spend the day in the woods cutting down trees. Then they would haul the trees to the sawmill and make boards. Sometimes, I would go along and play in the huge piles of sawdust. I could run up the big piles and then sllllliddee down! Daddy also ran a road grader and worked for Crawford County for a few years. When I was grown up, my Uncle Charlie built three or four big chicken houses out behind the fishing pond south of Roberta."

She describes the fishing pond so casually, as if I might go there tomorrow to catch a few fish myself. I wonder if the pond is even still there, or if it only lives on in her memory.

"Daddy took care of all them chickens for a few years. You would open the door and it was just a sea of white." She pauses, collecting her thoughts. "You see, Sierra, my Daddy never did have a great, important job, but he always worked."

"I see what you mean," I reply. My respect for T.O. is growing. The phrase he always worked echoes in my mind.

I ask, "What about the time your mama took you to Florida? Is there a story about that?"

She says, "When I was four years old, my mama come and kidnapped me. She and her brother acted like they were taking me out for ice cream. We got the ice cream, but then we headed out of town. I remember she'd make me duck down on the floorboard every time we saw a car. I think we got on a train in Atlanta and then took that into Florida. I have pictures of myself during that time, wearing a little white dress and my hair is cut short. I'm smiling for the camera with Mama Chris. While we were taking those photos, the state highway patrolmen were looking for me.

"I was a holy terror then. If I ever got out of the house, I would run away. One time, I saw some kids having a birth-day party in the backyard and I just walked right into the party. It looked fun!

"For a while, me and Mama moved in with her sister and husband. This was during the war years and everything was rationed. You could only buy things using the stamps allowed in your ration books. One time, Uncle Neil came home with a gallon of kerosene and a pound of sugar. Those were precious items. I don't know why but I took those things out in the backyard and dug a hole. I spent the afternoon making mud pies with the kerosene and sugar. Boy, was my backside sore when everyone found out what I'd done! I tell you, I was a holy terror.

"Anyway, they eventually found me and after a lawsuit, I went back to living with Daddy in Georgia. I was in Florida for about one year."

She tells this dramatic story casually, too, as if four-year-olds regularly got kidnapped by their divorced parents in the 1940s. Both Luther and Joan are so calm about the incidents from their childhoods, from Luther serving as a human gate in the back of a straight truck, to Joan's year in Florida. I wonder if I will ever shock my grandchildren with casual stories of life's dramatic moments.

Joan says, "Another time, I lived with Daddy until I was in the eighth grade."

Her Southern heritage shows when, as a seventy-some-year-old woman, she still calls her father "Daddy." She pronounces it "Dad-eee." Most of the young girls in Missouri outgrow "Daddy" and use the word "Dad" by age ten.

She says, "Then, I moved up to Missouri to be with Mama Chris in Centralia. By that time, she had married Kenneth Henry and lived in Missouri. After I moved, Aunt Belle stayed on for many years and Daddy kept living in that little three-room apartment. I lived in Missouri for three years, but when I was a senior in high school, I decided to go back to Georgia. In Centralia, the graduating class didn't do much special. But in Georgia, the senior class took a trip to Washington, D.C. and New York City. I thought that sounded like a fine trip. So, I decided to move back to Georgia for a year. After me and Luther got married, he would always go to the Centralia Alumni banquet each year. Technically, I didn't graduate from there, so I wasn't an alumna. However, I spent three years of my high school there so I knew everyone at the banquet. I was relegated to the non-alumnus section for a long time, but at least I got to go to Washington, D.C. with my senior class in Georgia."

I say, "I'm impressed that you moved away from all your friends just to go to Washington, D.C. for a few days."

Ignoring the compliment, she says plainly, "I guess I figured I might not make it there if I didn't take the chance while I had it. You have to understand, I grew up with those kids. I only went to school in Centralia for four years. The rest of my time was in Georgia."

I take the last bite of the bright yellow rum cake and scrape up the leftover bits of glaze from my plate. In the early morning, the heat from the liquor warms my mouth. I imagine leaving her house with rum-laced breath, so I reach for the bottle of red candies.

"Many years later," Joan continues, "after Luther and I got married, we had our first son, Jed. I waited until he was big enough to make the drive down to Georgia. I wanted to show him to my family. When I got there, the first thing I did was show T.O. his grandbaby. Then, I called Aunt Belle's daughter. I was going to take Jed out to her house to see Aunt Belle."

Joan pauses, a choke welling up in her throat. "I hate that I can't tell this story without crying. Damn, you'd think after all these years I wouldn't blubber so much about it."

She tries to speak again. "Anyway, Aunt Belle had gotten sick. She died in the hospital the day before." Joan is weepy now, and she smiles at me tenderly. Finally she says, "I never got to show Aunt Belle my baby. I was one day late. That was a real sad time."

I'm crying now, too, because what Joan doesn't say is: I never got to say goodbye to Aunt Belle. If Joan had arrived one day earlier, she could have tenderly held Aunt Belle's hand and said, "I'm sorry I jumped on your bed. Thank you for raising me. I love you."

Life sometimes gives us unfinished stories. But, had she been a few weeks earlier, I hope I know how this could have ended. Joan would have gotten a proper goodbye and a chance to say thank you. Aunt Belle and her daughter would have held the newborn baby, fussing and cuddling over his sweet, soft skin. As the maternal figure in Joan's life, Aunt Belle would have passed along her wisdom, "If he has colic try gripe water... if he gets a rash, use this...."

Although Joan wasn't able to say goodbye to Aunt Belle, she did get to stay in Georgia long enough to attend the funeral. She remembers that the family asked her to stand up and say a few words about Belle.

Joan says, "I'm not good at talking in front of people, but I said I'd speak. I got up and talked about how much it meant for Aunt Belle to raise me just like I was her own. At the end I said...." Joan's voice breaks off again. She sniffles and collects herself.

Joan explains, "I told the crowded church, 'If Aunt Belle didn't make it to heaven, I don't want to go.'"

Joan and Christine in Florida. Christine (McGee) Gassett Henry, 1917 – 1970.

T.O. Gassett, left, and the Sinclair Gasoline truck he drove. Thomas Olan "T.O." Gassett, 1906 – 1971.

T.O. at the chicken barn where he worked in Georgia.

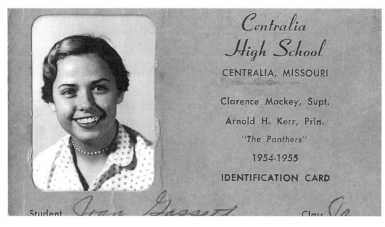

Joan's school identification card.

Joan poses in downtown Centralia, Missouri for a photo to send back to her family and friends in Georgia. Note the Vista Theater in the background.

Aunt Belle

Joan and Jed traveled to Knoxville, Georgia in 1961. They visited T.O., their extended family and Aunt Belle. T.O.'s older sister, Emidel "Memi" Havis of Decatur, holds Jed.

1953-1957
LUTHER'S FOUR YEARS
OF HOMESICK

CHAPTER 18

NAÏVE COUNTRY BOY

"In March of 1953," Luther says, "I was eighteen years old and going to the University of Missouri. I couldn't pass freshman English. After I flunked the class, they said they wanted me to take remedial English five days a week during the summer. That option sounded like pure torture to me and I thought just about anything would be better than taking that class. At eighteen years old, a fella is just coming to grips with himself and beginning to feel independent. Four of my friends had already joined the Air Force. I decided I wasn't going to take that awful remedial English class. Instead, I was going to run off and join the Air Force.

"I went to St. Louis to be sworn in, leaving from Jefferson City on a train. This was the first and easiest step of a journey that turned out to be the most challenging time in my life to that point. Now, I hadn't ever been too far from home.

I had never been out of the state of Missouri, nor had I ever went into the city before. I hadn't been much of anywhere really, but I was going to St. Louis to be sworn into the Air Force. I met another boy on the train from Fayette, which is another small town like Centralia. He was going to be sworn in and seemed to be nervous like me. They mailed both of us a map with directions pointing the way to the Warwick Hotel. We got off the train together. There we were, two naïve country boys scared to death in the bowels of downtown St. Louis. We finally found the hotel and we expected to safely share a room for the night. The hotel attendant directed us up to the fifth or sixth floor and we opened the door. There were already twelve people, all strangers, crammed into the room!

"The Warwick Hotel, I later found out, was nothin' but a cheap ol' flophouse. The Air Force was crowding new recruits into the rooms at the least possible expense. 'Flophouses' fell much lower on the scale of quality than a hotel or even a motel. Modern hostels, which are often considered 'roughing it' by teens backpacking across Europe, look like the Ritz when compared to our room that night. The Warwick Flophouse was filled with transient workers, plain ol' transients, drunks and, in the case of our room, fresh recruits. Instead of two beds, the room was packed full of bunk beds. We both found empty bunks and settled in. I can't say about him, but I didn't sleep a wink that night. Trying to sleep in that big, bad city was out of the question. I was just sure someone was going to knock me in the head and kill me.

"My first night away from home was memorable and scary. Yet, I'll never forget the next day either. There was supposed to be fourteen of us sworn in, but one of them chickened out and went home. So, there was thirteen of us sworn in on the thirteenth of the month—a Friday. That's right, thirteen of us on Friday the thirteenth! I am not superstitious but after a long, sleepless night of worrying about my premature demise, I thought it was an unlucky day to get sworn into the military.

"After that, they sent me and the other idiots home for the weekend. I had to go back on Monday. Oh, how I cried when I had to leave home for the second time!"

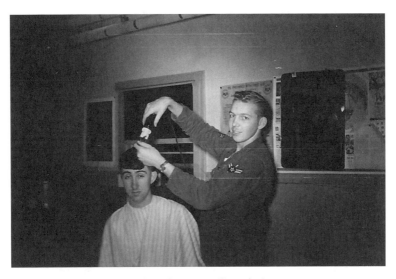

Luther is not enthusiastic about his new military haircut.

Luther and his high school girlfriend, Sharon Ward.

CHAPTER 19

LACKLAND AIR FORCE BASE
SAN ANTONIO, TEXAS

"On Monday, they put us on a train to the Lackland Air Force Base in San Antonio, Texas for basic training. On the train, there was only four or five of us country boys and the rest were slick city boys, mostly from Chicago. Although I don't remember most of them, two of the city boys I remember were the Post brothers. They decided to teach me how to play poker. I had about a hundred dollars in my pocket, probably more money than I had ever had in my life. By the time the Post brothers got done 'teaching' me to play poker, I was dead broke! I didn't even have a nickel to buy a Coke! This was the first of many lessons my new friends and experience in the Air Force taught me in the coming months.

"Since I 'spent' all of my money 'learning' to play poker, I had to wait thirty whole days until I got my first paycheck to buy a Coke. My first paycheck was eighty-two dollars. The first thing I bought was an ice cold Coke. I was raised on Coca-Cola. Those were the first weeks spent away from home in my whole life. It was awful! I was powerful lonesome and I was sure homesick like no person has been homesick before. That first Coke was the only familiar thing I'd seen, tasted, smelled or heard in a whole month. Never has a Coke tasted as good or meant so much to anyone else. If they were interested, I could have done an award-winning Coca-Cola endorsement.

"We were in Texas for what felt like a very long nine weeks. While I was there, my family came to visit for a few days and they brought my girlfriend, Sharon Ward, along, too. They stayed with our second cousins who lived nearby. After they left, I cried. The training at basic wasn't very hard, but I sure stayed home-sick. I'd never been away that long. I'd never been away at all!"

Luther and his mother, Louise.

Luther L. Angell Trains At Warren AFB

A/3c Luther L. Angell, son of L. W. Angell of R.F.D. 3, Centralia, is presently training as an Air Force Technician at the USAF Technical School at Warren Air Base, Wyoming, it has been announced by the Commanding Officer.

At this historic former cavalry post outside Cheyenne, the Air Force is training young airmen in the many specialities needed for air power. Besides clerical skills such as clerk-typist, clerk-stenographer, and administrative specialist, the schools at Warren Air Force Base, train automobile mechanics, powermen, electricians, teletype operators and repairmen.

e
le
le
s.
n-
r-
h
in

nt
vo
he
g
in

CHAPTER 20

FRANCIS E. WARREN AIR FORCE BASE
CHEYENNE, WYOMING

"When basic was over, they sent us to the Francis E. Warren Air Force Base in Cheyenne, Wyoming. This time we didn't take a train; they flew us. Of course, since I'd never been anywhere in my life, I'd never been on a plane either. So, that was my first plane ride in the Air Force. I was still homesick, but once we took off, I could add to that dreadful feeling. Now, I was feeling airsick, too.

   "The first morning in Cheyenne, I was asleep in the barracks. I woke up to the sound of a calf bawling. I thought I was having a dream, but I jumped out of bed and ran to the window anyway. Right outside of our building, there was a cattle drive going by. I guess they were ready to be loaded on the train and shipped east. I had not heard a bawling calf in over two months! I have never been so happy to see a herd of cattle in my life. The rest of the guys were still asleep, but I bet if they woke up they would have thought I was crazy.

"While I was in Cheyenne, they sent me to electronics school to learn about the new radar machines. I was about the dumbest son-of-a-buck there was, yet in all their wisdom the Air Force decided to send me on to further schooling to learn how to use the latest in radio and radar technology. Even when I graduated, I still didn't know nothing 'bout electronics. The machine they taught us to use was called a 'Philco'.

"At the time, this was pretty special technology; it had something to do with microwaves, and I'm not talking about the cooking kind. They were the type of radars used to detect missiles. Only a few of us in the whole Air Force got the schooling to run this brand new machine and somehow I was one of them. Or at least I was supposed to know how to run one. I am about the least mechanical person alive; I don't even change light bulbs or oil. So, I still do not know how I got roped into running those complex machines.

"Cheyenne was not a very big town at all, but while I was there, the world-famous Cheyenne Frontier Days Rodeo took place. It was a pretty wild time to be in town. The cowboys rode their horses right into the tavern and straight up to the bar. They never had to dismount to order a drink. I was not much of a cowboy then, so I was walking down the street taking in the sights when I heard somebody up the street yell, 'He---y Angell!'

"Just like the taste of that first Coca-Cola to my lips after my first service paycheck, the sound of a familiar voice was a welcome sound to my homesick ears. It had been a long time since anyone yelled my name in a friendly greeting. It turned out to be Toots Johnson, a local man, who was traveling with the carnival. I did not know Toots real well back home, but that didn't matter. He was hometown folk and I was glad to see him. It made me happy to hear someone say hello to me across the street. After we parted ways, seeing Toots actually made me more homesick. I wanted to see more familiar faces.

Luther's training group at the Wyoming base in October of 1953; Luther is in the middle row on the far left.

Luther completed 120 hours of training to learn how to operate the Philco radar machines.

My family came to visit again while I was in Cheyenne; that was real nice.

"I had been in the Air Force less than one year, but I already felt like this would be a long four years. I never had enough to do, so to pass the time, I started reading. Once I started reading, there was no stopping me. It was a little bit like when Forrest Gump decided to start running. He ran across the whole United States. During the five months I was in Cheyenne, I read the whole Bible. I'm not saying I absorbed every line in the Good Book, but I did read it. Reading helped pass the time, but at night I was still lonely and homesick. Sometimes, I'd lie in bed at night and wonder why I just hadn't taken remedial English instead!"

CHAPTER 21

SAN FRANCISCO, CALIFORNIA

"After Cheyenne, I went home for a few days before they sent me to my next assignment. Me, two other guys and my roommate L.C. Price headed toward Missouri together. I didn't have my own car, so four of us rode together. L.C. was a black boy from New Jersey and he was quite the ladies' man. On Friday nights, he would put on his 'go-to-hell' hat and the girls would be waiting at the gate for him. They just loved L.C.

"Occasionally, I had to work late on Friday night and we had room inspections early every Saturday morning. When I was working late shift, L.C. would straighten up my side of the room. Then he made sure my shoes were shined. He never went out before everything was ready for inspections for both of us. He was one of the nicest guys I ever met.

"We stopped in Kansas City to get some lunch on our trip

back home. As we walked into the restaurant, everyone sort of turned and looked at us funny.

"They said, 'We can't serve you all.'

"This was during the early 1950s, so most of the country was still segregated. We had to leave and find another place for lunch. We weren't surprised; it was just the way things were back then.

"I made it home and then I waited to find out where they would send me next. When the order came through, I was not excited about my new location: Fire Island, Alaska. I took a train all the way from Centralia to San Francisco. From there, we traveled by boat to Alaska.

"Luckily, I had something to look forward to when I got to San Francisco. My cousin Eddy Chamberlain was in the service there and he was going to meet me at the train station.

"Although I was bound for Alaska, Eddy was stationed in California. I thought he was pretty lucky. He got the beach and I got the frozen tundra. I sure drew the short straw on that deal! Growing up, Eddy learned to cook at the Globe Hotel in Centralia. He became a very good cook and everyone was sad to see him leave the hotel for the military. During basic training, the officers figured out that he was an experienced cook and took him right out of training and into the kitchen. He spent his entire four years in the service cooking in one mess hall for the same set of officers out in California. I think every time Eddy was supposed to be assigned to a new location, those officers would make sure his assignment changed so he stayed at their base. They liked his cooking so much, they never let him leave.

"When I finally arrived, it turned out to be a good thing Eddy was meeting me. The darn train lost my bag somewhere between Missouri and California. Eddy knew the San Francisco area pretty well, so he took me down to the riverfront a few days later. Somehow, he found my bag at the shipping docks. I never would have gone to the shady part of town without Eddy. It was rough and I was glad when we found my bag and got the hell out of there.

"A few days later, I was on a boat heading toward Alaska. We sailed right under the Golden Gate Bridge. Boy, was it was beautiful. The sun was shining and I was feeling pretty good about things and thankful that Eddy had found my bag.

"Then, our ship broke down. We were stuck out in the middle of the ocean for a couple days. We swayed back and forth at the mercy of the waves waiting until the boat was fixed. It was impossible to sleep, because we got seasick from the swaying. There were three or four bunks stacked on top of each other to conserve precious space on the ship. The next mattress was about six inches above my head. Things got real bad when everyone started puking over the side of their bunk. I kept my face covered and stayed in the center of my small bed, hoping not to get hit by vomit. My time in the military was just making me sick! First I was homesick, then I got airsick and now I was seasick, too!

"Eventually, they got the boat fixed and we started toward Alaska again. It took a few days for my stomach to settle down. Once we made it to the coast, we took a train inland to Anchorage. Then a small plane made the final leg of our journey to Fire Island. All of this made for a long trip, but the breakdown at the beginning was the worst part."

Luther mailed this photo home to his folks with the caption, "Yours truly, preparing for war."

CHAPTER 22

FIRE ISLAND, ALASKA

"We arrived at Fire Island just two days before Christmas in 1953. Fire Island is located in the Cook Inlet. Basically, the island was an old volcano. It was only seven miles long by two miles wide. It was my first Christmas away from home and I didn't have any friends yet, so I was mighty lonesome. My roommate was new, too. His name was Chico Mandez. Chico was a California Mexican. I was a plain old country boy and he was a smooth talking city slicker. Obviously, we did not have much in common at first. Yet we only had each other, so the two of us stayed in our room all night long. I had a record player, but I only had one record. We played it so many times, I still remember every word to the songs on the record. It was by Kay Star and the Wheel of Fortune, only two songs on each side.

"I can hear her singing those four songs, with the record

Chico Mandez

spinning round and round, round and round. Me and ol'
Chico laid in our bunks looking at the ceiling all night on
Christmas Eve. Damn, was I ever homesick those first few
nights in Alaska!

"After I got my first paycheck, I bought a bottle of Chanel
No. 5 perfume and mailed it home to my girlfriend. Sadly,

she wrote back to say the bottle was empty when it arrived. I guess I didn't get the lid back on tight after I smelled it. I was bummed that I wasted my money. Things just were not going my way.

"Before January, I figured out life that on Fire Island was going to be dull. I doubled down on my reading, and I made it through several books each week. The Air Force assigned two hundred soldiers to the base. The way I figured, about forty or fifty men could have handled the work allotted to our entire base. Needless to say, at any given time, about a hundred of us had nothing to do.

"Fire Island was a military radar base. I happened to be on base at the height of U.S. fear and tension over the threat of nuclear war with Russia. We were using the high-tech microwave machines to keep an eye on our 'neighbors' - the Russians. From the sounds of the political situation, we should have been busy and stressed while monitoring the impending nuclear warheads. However, my experience was just the opposite. Most of my memories from Fire Island have nothing to do with military work or stress. We had too many people for too little work. Mostly, I remember spending my time trying doing things to keep me from being bored.

"In the spring, me and a friend explored every single inch of the island. We knew every pond, creek and cliff. When it was warm, the island was a pretty nice place to live. It was even hot enough to swim. In the winter, it could be forty or fifty below. Those days were awful. An interesting thing about living on base in Alaska was the wooden tunnel system connecting all of the buildings. This way we didn't have to go outside when it was wintertime. The tunnels were 150 to 170 inches high. Occasionally, we shoveled them out a little bit to keep them clean.

"One good thing about life on the base was the unlimited supply of booze. I would have rather had an unlimited supply of Coke. I could not get any Coke on the island. I wrote home to tell my mama about the Coke shortage and she started

137

sending them to me in the mail. When the Cokes arrived, we tied them on the end of fishing line and dropped them down in the ponds so they could get cold. They were delicious.

"We had a softball league to help keep us busy. I did not care much for softball; I played anyway just to pass the time. From the base, I could see Anchorage. The lights reminded me of the world off Fire Island. During that time, Alaska was in a kind of a frontier stage like the Old West. I bet there were a hundred bars and two grocery stores in every town. The city was close to our base, but we never got to go. Instead, I was stuck with two hundred restless soldiers and not one single woman. From December of '53 until April of '55, I was stuck on the island without any women. Those were the longest, most tedious months of my life.

"I guess there was one woman, the librarian. She was the highlight of the whole damn thing. Once a month, she flew to the island and brought us new books and took back the others. You should have seen how two hundred female-deprived soldiers would line up for books when a woman was passing them out. We would have stood in line for a bag of rocks just to see her. I bet most of the guys did not even read the books they checked out. The librarian must have been at least sixty years old, but we did not care about her age. Hell, she was the only woman we got to see.

"Of course, I lined up with the rest of the guys to get books. I think I was one of the only ones that actually read the books. Reading is what I did more of than anything while I was in Alaska. After I finished the Bible in Cheyenne, nothing seemed long or intimidating. I read all kinds of books— short books, long books, classic books, mystery books. It was easy for me to read all day when it was forty degrees below outside and I had nothing else to do.

"I worked in a little building that we called the Radar Room. I worked my shifts in the room with all the screens and knobs and I would often fall asleep. Now I know, most people frown upon sleeping on the job. Especially when

Luther in the radar room.

you're supposed to be monitoring something like Russia's nuclear bombs. The problem was, with my job, there wasn't anything to do except wait for something to go wrong. To make sure everything was going smoothly, I just had to do a quick check once every hour. Between my checks, I sat and stared at a wall full of screens, buttons and dials. Nobody could stay awake for that.

"At first, I was a little nervous about getting caught sleeping on the job. Yet, I quickly realized there was no way to get caught because of the layout of the building. To get to the control room and check on me, a commanding officer had to come through two doors. When he came through the first one, it would slam and wake me up. By the time he got through the door into the Radar Room, I was sitting upright in my chair dutifully doing my job.

"One of the other tasks we had on Fire Island was kitchen police. We called it 'pulling KP'. A lot of the guys didn't want to work their shifts, because it meant eight hours of cooking or doing dishes. Instead, they would pay ten bucks to whoever wanted to work for them. I didn't really mind it, because there was nothing else to do on the island. Pulling KP helped me pass the time and I liked making a little more money. I always sent home my Air Force paycheck along with the KP money.

"Pulling KP wasn't a glamorous job though, there were reasons some guys didn't like doing it. The food in Alaska wasn't exactly farm fresh. The Air Force sent us a barge full of food every summer and that was it for the year. We stored all the food in a warehouse. In Alaska it gets above freezing during the day and then drops back down again at night. All the food would freeze and thaw day after day in the big warehouse. After months of this the taste and texture of our food changed dramatically.

"Years ago, some officer must have made a mistake when they ordered the eggs for Fire Island. They must have added a few zeros to the order. Sometimes, I made breakfast for everybody on base using eggs that expired ten years before! Nobody got sick from them so we just kept trying to use them all up.

"Old or new, it takes a lot of eggs to feed two hundred men breakfast. I had a huge silver bowl to crack eggs into. The job would take at least an hour. Every once in a while, I'd pick up an egg, crack it and a little chicky would fall out and sink into the big bowl. At first, I'd reach in there with a spoon and scoop out the chick. I tossed it into the trashcan and went back to cracking more eggs. To a 'civilian', I realize this experience in itself would be shocking. Imagine if that happened at the local Hilton Hotel? They would probably be shut down within a day.

"Yet we weren't at the Hilton and the egg situation actually gets worse. One day, I saw another guy cracking eggs for the base. He didn't scoop out the little chicks like I did. Instead, he took his bowl over to the mixer. He blended up the egg yokes and them little chickies all together. Nobody said a word about his odd procedure and no one ever noticed. After that day, I stopped scooping 'em out, too!

"They also sent us powdered eggs, even though we still had the expired ones. For some chemical reason unknown to me, the result of the freezing and thawing turned the egg powder green. It still tasted the same, but it was a washed-out shade of green. It was sort of like Dr. Seuss's "Green Eggs and Ham" book. Let me tell you, it wasn't appetizing in the book and it was even worse in real life.

"Since our food came by barge, the flour delivered in the spring was full of bugs by the fall. It was our only supply for the year, so we kept making bread with the bug-infested flour. At first, me and the other new guys would pick the bugs out of our bread. Then, after a few minutes of work, we would eat our bread. The guys who had been on the island would laugh at our pickiness and eat the bugs right along with their bread. Before long, we were doing it, too. It might sound gross, but I honestly could not tell a difference with or without the bugs.

"The second winter I was up there, it snowed 130 inches. It was 1954. In the summertime, Fire Island was a nice place to be, but during the winter it was awful cold. Sometimes they would make us practice war games. We'd go outside and pitch a tent when it was forty degrees below zero. We had to sleep out in the wilderness for a few nights and pitch a fire. We were only fifteen miles away from the Russians. I guess those war games were supposed to help us be prepared for a Russian attack or something. Mostly, I just thought it was too damn cold to be camping out.

"Other than that, I had a lot of free time. I made friends with the guys on base and we did our best to keep from being bored. Sometimes we would play cards for twenty-four hours in a row. I was only making eighty-five dollars a month, but I always sent home more than $140. Between the cards and KP, I could usually make another sixty dollars. We played double-deck pinochle, poker and sometimes hearts. I got real good at poker, unlike my first ride to boot camp when I lost all my money.

"Me and my roommate Chico also became good buddies. He had a guitar. Since there wasn't anything else to do, he decided to teach me how to play. Chico was real patient with me, but I was just awful. I could hold the pick in my hand and strum along the strings, but when it came to the cords, I was flat unteachable. Chico and I spent months working on it, but I just couldn't get my fingers to move and remember where to land on the little strings. We finally gave up. It was to no discredit to ol' Chico; he sure tried.

"Another thing I did to pass the time was paint by number pictures. I wasn't much for drawing and I've never been a real artistic person. However, I could fill in the little holes with whatever color the numbers told me to use. It was easy and I loved doing those paintings. It was fun for me to watch all those little spots of color become a real picture. It might be not a very manly thing to say, but I don't care. It was neat.

"One of the pictures I did was of a naked lady. Me and Chico liked that one. I tacked it right up above my bed. One day, Chico took my photograph while I was lying on the bed. A bit later, I decided to send the photo of me home to my mama. I wrote about having plenty of down time on the island. I thought the picture of me sprawled out on the bed went well with my note. When she got my letter, she wrote back that she wasn't too happy with me. She was not impressed with the choice of artwork hanging above my bed. When I sent the picture home, I completely forgot about the naked lady painting was right above my head. After I saw the picture again, I understood why Mama was mad.

"Chico never did do any paint by numbers, but he did write lots of letters. He would sit at his desk and write letters to all his girlfriends. Chico sure liked the ladies. I guess he decided with no women on the island, he was going to have lots when he got home. I never really got a good count of how many women he wrote to. He was pretty deliberate about keeping up with all of his girlfriends; he took three or four sheets of paper and wrote one sentence on each page. When he would go to the next letter, he wrote the same sentence over again. He wrote the same thing to each girl so he didn't have to keep track of what he said. Every letter was the same. Except that each letter was signed from a different name like Chico, Ace or Jose. I guess Chico thought using different first names made sure that if the girls knew each other, they wouldn't figure out what was really going on.

"Me and ol' Chico did pretty well in Alaska, despite the poor food and boredom. By the time the second Christmas came around, we both made lots of friends. One of my best friends was a superintendent's son from Cimarron, Kansas. The base had a little bar with about three our four tables; the whole place was about as big as a living room. That second Christmas was the first time I got soused on rum and Coke. That was a Christmas Eve to remember. Every month, the Air Force sent us a small plane full of booze. During the summer, they sent Lucky Lager beer by the barge. It got pretty rank by fall, but we drank it anyway. The little bar on base had a jukebox, but it only had twelve songs. One of them was Doris Day's 'Sentimental Journey'. At night when it was really clear, we could pick up the radio station out of California. Their program was called 'Lucky Lager Dance Time'. Some folks didn't fare as well on Fire Island. The lack of civilization made them crazier than a pet coon. I swear one of my captains up there thought he was Billy the Kid. He had been isolated for too long. He walked around wearing pistols on his hips. Our medic was crazy, too. I'd rather treat myself than go to his loony bin for medical advice. On payday, we would all go down to the Post Exchange, or the PX, and collect our checks. This started off the drinking for some guys. Before long, they had spent their whole paycheck, but still needed a buzz.

"Then, when they got desperate, they drained anti-freeze out of the trucks and strained the alcohol out through bread. That was supposed to take the bad stuff out, but a few times, it did make them blind. It even killed a few people. Another guy I knew would drink away his paycheck and then start in on the Aqua Velva after-shave. He took his sleeping bag out into the woods for a few days at a time with that after-shave. I don't think Fire Island made that guy lose it; he was half crazy when he arrived.

"The highlight of my time on Fire Island was when they brought Roy Acuff all the way up there to sing for us in a USO show. The officers wouldn't let us have any booze during the show, but it was still really fun. When it was finally time for me to leave Fire Island, the Air Force was supposed to send a small plane to fetch me and a few other guys. It didn't come. After a day or two of waiting, we hired our own plane. We were ready to get the hell out of there no matter what it cost! The monthly load of booze came in by air and we got on for the backhaul.

"Years after I left Fire Island, it became the first place in

Alaska where they discovered oil. When I got home on leave, I found out that my girlfriend had deserted me. After two long winters without seeing any women, I sure was looking forward to her company. The first night, I was playing basketball at the park with my friends and I broke my foot. Obviously, it wasn't a good trip home after those two things happened. Damn these four years!"

Luther mailed this photo home to his mother. She was not pleased with the artwork above his bed.

Luther typed this caption onto the top of the photo: "Yours truly, working as usual."

CHAPTER 23

MOUNT MITCHELL, NORTH CAROLINA

Luther says, "I was assigned to a base on the top of Mount Mitchell, but it wasn't much of a base. It was really just a cabin and a small radar station. There were only eight of us stationed there. Mount Mitchell is the highest peak in the Appalachian Mountains and in the eastern United States. The mountain is big enough that it is located in two states. The top, where I lived, is located in North Carolina. When we drove the winding, twenty-eight-mile road down to the bottom, we crossed into South Carolina.

"Our main job was to keep four of the eight generators running at the radar station. To do that, we had to haul fuel. At the bottom of the mountain there was a little tiny town called Garden City. It was sorta like Thompson, Missouri before the Dawson store closed down. What I mean is, this little town had nothing but a gas station and a couple houses.

We had a military credit card, so we got fuel in Garden City and then headed up the mountain.

"In the summer, we could drive almost all the way to the top. The last mile and half before the cabin, there was no road. We hauled the fuel by foot. There was also a water line up the mountain to the cabin. In the summer, it delivered enough water for us to drink and cook. However, there wasn't enough to shower. In the wintertime, the line froze because it just ran above ground.

"This was one thing I didn't like about my job there on the radar base. I think the military tried to send me to the most trying places on earth. First, I had to endure two winters in Alaska. Then, they assigned me to haul fuel and water to the highest point in the eastern United States! Some soldiers do meaningful work in the military, but not me! I swear they gave me the most meaningless jobs they could find. I wished I was hauling water to some cows, instead of to the top of a damn mountain! I sure felt like one unlucky son-of-a-buck some of those winter days. I trudged upward, fuel can in each hand, cursing the fact that I was literally climbing the highest damn mountain around for a thousand miles.

"Our cabin didn't have any showers and little Garden City didn't have a place for us to shower either. We went up the road to another town called Marion. There was a barbershop there and the barber let us shower there for a quarter. We called his shop the 'greasy corner'. After several days of mountain living, we were sure glad to get cleaned up with a hot shower.

"In Marion, I met a pretty girl named Nancy. Her dad ran an auto body shop in town and I was sweet on her. She was purdier than hell. Even today, my wife lets me keep a picture of pretty little Nancy. While I was dating her, all we ever got to do was go to church. Her mama was awfully conservative, so we weren't allow to go on many dates. Her family went to church several times a week. I figured out pretty quick, if I was going to see Nancy, I had to go to church. Boy, did I spend a lot of time in church. We went on Sunday mornings and then back on Sunday nights. Then on Wednesdays, we went again for the prayer meeting.

"Me and Nancy went to a picture show a few times, but mostly we just sat in church. I was pretty well-behaved at that stage in my life, but I guess that's because I spent so much time at church I didn't have a chance to go looking for trouble.

"The folks in Marion were real good to us. We were the only eight military men around and they treated us like we was the best of the best. I didn't think hauling fuel and water was a very special job but in their minds, we were servicemen. Because of that, we were well-respected. Some of the locals would have done anything to help us out.

"One time, we were filling up our fuel cans to haul up the mountain. There was a little park ranger station next to the filling station. That day, a park ranger came over to ask for our help. Of all things, he needed help delivering a baby bear. I don't know a dang thing about bears, but we helped out best we could and the little bear cub survived. That was pretty big news in a small town. After that, we were really treated like heroes.

"Another time, we were down in the valley and a snowstorm blew in unexpectedly. All the motels filled up, so we didn't have anywhere to stay for the night. We decided to go to the local jail. He let us spend the night there until the storm passed and we could get back up to our cabin. I can honestly say this was the only night of my life that I spent in a jail cell. Plus, this is probably the only situation where I could say that staying the night in a jail wasn't a bad thing.

"If I wasn't at church with Nancy or hauling fuel up the side of the mountain, I usually loafed at Mouse Mitchell's general store and gas station. Mouse and his friends would sit around drinking coffee and swapping stories all day. They all called me 'Catfish' because of the stories I told them about Missouri's big catfish.

"I told them, 'Back home, a person could fill the back of a pickup truck with big ol' catfish out of the Missouri River.'

"They thought I was lying, but I wasn't. That is the damn truth. Mouse had a little filling station with his store. Back then, the filling station managers would come out and fill up your tank for you. This service was always free of charge. You could even ask them to check your oil. Them older cars used a lot more oil than our modern ones.

"One day, I noticed that Mouse always took a quart of oil with him when he went to check the oil in someone's car. I asked him about it and he said, 'If I am going to check the oil, they are going to need at least one quart.'

"A lot of the locals were funny characters like that. Another one who liked to loaf at the filling station was the county sheriff. He looked out for us because we were from out of town. While I was living there, they were in the process of desegregating. Things were a lot different than you might imagine. First of all, there was hardly nothing around to desegregate. People were stuck in their ways, too.

"Back then, some of the black people would walk up and down the roads at night in groups of fifteen or twenty. The sheriff told us, 'If you are ever out driving at night and you hit somebody, do not stop. If you stop, they will kill you. You just keep driving. You come get me and then I will take care of it.'

"That's just the way it was in the Deep South back then. I never had any trouble, and I'm sure glad I didn't. I really came to enjoy that little cabin, Marion and Garden City. Many years later, I took my wife there to visit. I was so excited to show her the cabin and the radar station. When we got to the top of the mountain, it was gone. I got choked up when I saw the empty spot where the cabin used to stand. We drove an awful long way from Missouri to see that cabin. I turned our car around and we headed back down the mountain in silence. I should have known that a little ol' cabin wouldn't be sitting in the same spot thirty years later. But I hoped it would be there."

Luther's mountaintop cabin on Mount Mitchell.

CHAPTER 24

CENTRALIA, MISSOURI

During his time in the military, Luther traveled to Texas, Wyoming, California, Alaska and both the Carolinas. He also went to a few other states on short dispatches. He met lots of people from lots of different places – like L.C. from New Jersey, Chico from California and Nancy from North Carolina. By the time Luther finished his four years in the military, he had changed. He simply recalls being homesick, but much more happened in the formation of his character during that time. Luther left Centralia a green, naïve country boy and returned a poker-playing, well-read young man.

Luther says, "I was homesick the whole time. Well, except when I was loafing at Mouse Mitchell's or in church with Nancy. When I was discharged, there were only thirty or forty people in the whole country who had graduated from the Philco Microwave Electronics School. I was considered an electronic specialist. Nobody was more surprised by that certification than me. There were only a handful with that certification and I could have made lots of money in the private sector. But I didn't want to work on electronics."

When Luther finished his time in the service, he knew exactly what he was going to do for the rest of his life. He wanted to work at the livestock auction, buy livestock and live in Centralia.

Luther says, "Some people say that the happiest day of their life was when they got married or when they had their first child. Those were pretty happy days for me too, but they don't rank at the top of my list. The happiest day of my life was in February of 1957 when I drove down the main street of Centralia with discharge papers in my hand. As I looked out the window and passed by the familiar buildings, houses and people, I thought to myself, 'I'm home! I'm home! And, I am never gonna leave again!'"

1957-1959
A WEAKNESS FOR WOMEN

A model replica of Luther's 1957 Chevrolet convertible.

CHAPTER 25

BIG UGLY

"One of the first things I did after I got home from the military was buy myself a 1957 Chevrolet convertible," Luther says. "The exterior was black and white. It had whitewall tires and the inside had red leather seats. It was the finest looking car you ever saw. Woooo-wee! I just loved riding around in that car."

The way Luther saw things, he had just wasted nearly four years of his life sitting around waiting for things to happen. Those Air Force bases were boring and he was ready to start living at a faster pace. The convertible helped set the tone for his new outlook. He also enrolled in college again, hoping to finish the degree he'd started four year ago.

Luther says, "I moved into a house in Columbia with several other guys from Centralia. Within my first semester, I passed English 1000. I was recovering from those cold, lonely

Luther looked at this photo and said, "Ain't I a handsome dude!"

months in Alaska by going on lots of dates. Some of them turned out pretty good, and others were pretty bad. I'm just gonna tell you about a couple of my worst dates."

As a teen, Luther was pimpled, painfully scrawny and shy. He preferred to stay home on Friday nights reading books. But four years in the military were good for his complexion and build. Although he remained a thin 135 pounds, he had finally filled out a little. He wore size 31x31 slacks, his arms were toned and he looked like a man rather than a gangly teen. After he was discharged, Luther continued wearing the short, cropped hairstyle that the military instituted. It suited his dark hair and the times; he discovered that college girls liked the look of a mature servicemen. Although his time in the service was far from heroic, Luther was happy to play the role of "military man." Rather than feeling awkward around girls, he now had stories of faraway places like California, Alaska and South Carolina. His pointed-toe Justin cowboy boots and a last name pronounced "angel" added to his allure. Much to his surprise, Luther had become something of a ladies' man.

"That fall, I met a girl from the Christian College in Columbia," Luther says. "I thought she was pretty cute, so I asked her out. The night of our date, I went to her dorm to pick her up. She needed to sign out with the dorm monitor when we left and sign back in when I took her home. The college dorms were strict about the girls going out on dates back then. At the end of the night, we rounded the corner and started walking up the sidewalk. We were going to sign back in at the front desk. She stopped suddenly and said, 'Oh my God, that's my boyfriend!' I didn't even know she had a boyfriend. He looked big and I knew we needed to get out of there quick."

They headed around the back of the dormitory, hoping he had not seen them. The couple ran in the dark, dodging trees and trying not to trip over fallen branches.

"I decided the fire escape was my chance to get this girl home!" Luther says.

He jumped up and pulled the down the ladder, but it jammed and didn't come all the way down to the ground.

"I grabbed her by the waist and lifted her up," he says. "She grasped onto the bottom stair and tried to pull herself up, but she wasn't strong enough. I started thinking, this gal is a real pain. First she doesn't tell me about her boyfriend and now I have to save her? I started pushing on her hind end. About the time I was really giving her bottom a good ol' squeeze and shove upward, her boyfriend came around the corner. I hoped he didn't see me squeezing on his girlfriend's behind!"

Luther's voice gets louder and louder. "She was halfway up the stairs and I figured it was high time for me to get out of there. I left her with her feet dangling from the fire escape. I might have felt bad for her, but this whole situation was her fault. I didn't know about her boyfriend. I hurdled a row of bushes and made it to the side street where my convertible was parked. Her boyfriend never saw my face and he never did get a hold of me, I just sped away. As I drove down College Avenue, I decided this bad date had a few upsides: she got home safe and I had a good story to tell my friends. When I made it back to our house on Walnut Street, I told them about my big adventure. To make things real dramatic, I started calling her boyfriend "Big Ugly."

"A few weeks later, me and my roommates were playing cards all night long. My brother Buddy was playing with us, too. He had already lost all of his money, so we said it was his job to go downtown and buy donuts. We gave him some cash and sent him out. His car was blocked in the driveway, so instead he took my convertible. Just one street over from our house, there was a really good bakery. The donut shop was on Broadway Street, so they called it the Broadway Bakery. Buddy got the donuts and started to head back to the house. When he pulled away, he noticed a car was following within a couple inches of his bumper. As he turned back onto our street he noticed the same car was still following him."

Buddy started to get nervous, and felt unsure of what to

do about the car. He decided maybe he had a taillight out, so he pulled over and got out. When he stepped out of the car, two big guys met him at the door. He could barely see their faces in the dark street.

Luther says, "One of the guys grabbed Buddy and threw him right on top of the canvas top on my convertible. He said, 'You think you gotta go to the Christian College to find a girl? We catch you over there again and this is gonna get a whole lot worse.' By the time Buddy made it safely back to our house, he still looked white as a ghost. I could tell right away he was upset with me. Buddy said, 'Two guys just threw me on top of your car! They didn't hurt me, but it sure did scare the devil out of me. They were really big guys.'"

Luther knew it was Big Ugly, so he hemmed and hawed around an answer, but one of his buddies answered for him. "Did Luther tell you about Big Ugly? Last weekend, he took somebody's girlfriend out on a date!"

"What?" Buddy asked, growing more angry.

"Buddy," Luther pleaded, "you don't understand! I didn't know she had a boyfriend – she didn't tell me 'til we about ran right into him! I had to send her up the fire escape because he was blocking the door. He must have seen my car when I drove away."

"You are always into something!" Buddy said. "I should have known to never take your car anywhere. I'm never driving that car again!"

Luther never did run into the two big fellows again, but he did figure out who "Big Ugly" was.

Danny LaRose was one of the University of Missouri's greatest two-way talents. In 1992, he was inducted into the University of Missouri Hall of Fame. He played football for MU as #87 and threw the shot put for the track team. Around the time that Buddy and Luther had their run-in with Danny, he was selected as the "Big 8 Conference Sophomore of the Year." He also held the indoor shot put record at the University of Missouri for many years. Buddy wasn't kidding about getting tossed onto the roof of Luther's car. He probably seemed pretty light to a defensive lineman who was also a record-holding shot put thrower.

Buddy and Luther went on to stay in the Columbia area, selling livestock – but Danny moved on to a professional career with the Detroit Lions, Pittsburgh Steelers and Miami Dolphins. No one is really sure where the girlfriend ended up!

CHAPTER 26

A BLIND DATE

Luther sneaks up behind Joan while she stands over the kitchen stove. She is preparing fat hamburgers in a heavy cast iron skillet. He hugs her and says in a silly voice, "Mama, I loooooveeeee you. Have I told you that today?"

Then, he plants six short, loud kisses on her neck. She giggles and shoos him away playfully.

"Have I told you about the time I stood your grandmother up?" Luther says while giving Joan a goofy, apologetic grin.

Luther begins eagerly, "You see, I was supposed to go on a blind date one afternoon. And that was all I knew. I was excited to spend the afternoon sporting a pretty model at the University of Missouri football game. Later that evening, I was supposed to take your grandmother here out for supper."

Luther's military man look was earning him plenty of dates, but occasionally he still let a friend set him up on a blind date—especially when he was told that the girl was a professional model. Many months without women on the frozen tundra of Fire Island pushed Luther to a new level. Two dates in one day was only making up for lost opportunities.

He says, "It was a beautiful early September day. It was about ninety degrees outside and the sun was shining. When it was time, I went to the house to pick up my date. I wasn't sure how she would look, but I had high hopes. After all, she was a model! When she stepped out on the porch, I knew right away she wasn't the good-looking gal my friend had promised me. This gal was plain. In fact, it was worse than that. She was pretty hard-looking. I'm not the cream of the crop either, but I don't go around getting people's hopes up telling them I'm a model. Boy, was I let down! On top of that, here we were in the middle of a hot day and she was wearing an ankle-length mink coat and white gloves. I was damn near embarrassed by her get-up. And, if you know me at all, you know I don't get embarrassed. This outfit was bad. It turned out this gal was a hand model."

Luther pretends to pose, modeling his hands and arms for an invisible photographer. The burgers sizzle and pop while Joan slowly cooks them. The kitchen is designed so that the stovetop is located in the center island, next to the kitchen table. This way, Joan can stand – spatula in hand – preparing food and talking with Luther and I. Unlike most homes, standing over the stove doesn't force her to look at a wall while everyone else talks around a table.

Above her stovetop, there is a pot and pan rack hanging from the ceiling. Twenty pots and pans, a variety of stainless steel, cast iron and non-stick, hang prepared for convenient access.

Luther says, "It was too late to avoid our date. Sometimes when you're in a situation like this, I learned that drinking a little will make things look better. I always took Purple Passion in gallon jugs to the Missouri football games. It is a mixture of grape juice and vodka with a little splash of lemon. I started drinking, hoping that a bit of Purple Passion would help her looks and make me feel better about my situation."

I roll my eyes; hearing my grandfather talking about his tipsy football games is a bit ridiculous at times.

"Despite my enthusiastic attempt to make this gal more appealing, by half-time she still didn't look pretty. So, I doubled up." He says with a chuckle, "Damn it, this was the first gal I had ever been with that I couldn't drink away the ugly! It wouldn't have been so bad if she turned out to be nice. Instead, she was painfully vain. She told me, 'I've got to protect my hands from the sun.' She wouldn't take those darn white gloves off the whole game. By the end of the game, my friends had to drive us both home and they put me right to bed. I had too much Purple Passion and I was feeling pretty bad. Now that night, I was supposed to have a date with your grandmother. The game was in the afternoon and I was going to take her to supper. By the time I woke up, it was too late for my second date. It was long, long past when I was supposed to pick up Joan. I figured it was a lost cause—I didn't even try to call her to apologize!"

Luther looks guiltily over at his wife, "That's right, isn't it, honey? You were pretty upset with me, weren't you?"

Joan nods her head yes. Luther says, "Don't worry, sweetie, I ended up with you. Now I tell people about the night I accidentally stood my wife up for a model. Nobody has to know she was the sorriest-looking model I ever saw."

CHAPTER 27

HIPPIE GIRLS AND STEPHENS GIRLS

Luther says, "Another one of my college friends was L—. He was a married man, but he had a weakness for women. He shoulda been at home instead of out running around with me in his fancy car, but we sure had fun together. We always mixed up the same drink for our nights on the town. He'd buy a gallon of apple cider and a fifth of vodka. Then he dumped out just enough cider to make room for the vodka. Hmmmmmm-mmmm, it was a good tasting drink! One evening, we took his car out driving in downtown Columbia. Before long, we met two hippie girls. We'd been working pretty hard on our apple cider jug and somehow we talked the two hippies into riding around with us."

Luther and L— were excited to have two ladies to join them. The girls jumped into the back seat and L— sped away.

"These were the first two hippies I ever saw in my

whole life," Luther says with excitement. "Boy, they were strange-looking ladies. Both girls wore long, black skirts. Once they got in the car, I noticed they had dirty, long hair. They told us they were attending Stephens College, but they were originally from New York City. Since Big Ugly had run me off from the Christian College, I figured I might as well hang out with these two hippie girls from Stephens College. The dorms at Stephens College had a curfew, too. By the time we got the girls back to their dorm, it was way past the time when male visitors were allowed inside. I should have just followed the rules and gone home. But, one of them hippies had an idea. She said, 'Why don't you sneak in? We live in the room right over there on the ground floor. We'll go in and open the window for you.'

"I thought that sounded pretty fun! L— drove me around back and I went to go look for their room, but the gates were locked. That wasn't going to stop me! Over the top I went."

Luther began pulling himself slowly up the wrought iron fence. His movements were slow and clumsy. The apple cider jug was now empty. As he threw one long leg over the fence, he lost his balance.

"I fell right in the middle of a damn rose bush," he yells. "It hurt like you wouldn't believe. It took me a few minutes to get untangled from that bush, but I was still ready to find the room. I made it to the window she told me to find."

Luther knocked quietly on the window. Nothing happened, so he started knocking on all the windows nearby, too.

"Nobody opened a single window," he says. "I finally gave up. I climbed back over the top of the fence and I made back to the car. I figured those girls lied to me about which window belonged to them. When I got to the car, L— said, 'Man, you look like you been in a war! It looks like a damn bobcat got a hold of you!' I looked down and my shirt was all bloody and dirty from my fall into the rose bush. I ripped my pants on the fence and leaves were sticking out of my hair. I

172

guess I did look pretty sorry. Maybe those two hippies saw how bloody and dirty I was and didn't want me to come in anymore!

"It seemed like me and L— was always having a good time, and running right on the border of getting into serious trouble. L— lived in small house next to a big, fancy house out in Columbia. Instead of paying rent, he took care of the big house. The people who lived in the larger home managed the University of Missouri bookstore. They had a teenage daughter who wanted to go to Jefferson City to a football game. She wasn't old enough to drive, so she asked L— to drive her down there. I went with them and we dropped the girl and her friend off at the game.

"We took advantage of being in Jefferson City for the evening by visiting the Rathskellar. It was the best bar in town. We were supposed to pick the girls up around 9:00 or 9:30 after the game. But we forgot all about those poor girls once we got to the bar and started dancing and drinking. By the time we finally remembered the girls, it was 10:30 at night.

"We made it back to the football stadium and those poor girls were the only ones left at the field. They were huddled together under a big light pole. L— thought he'd really done himself in that time. He figured that she would tell her folks and he'd be in hot water with his landlords. I have no idea why, but she never said a word! L— always seemed to find his way out of trouble!

"Another one of L—'s jobs was driving the school bus for Stephens College. It was his job to pick up the girls and drive them out to the horse barn for their riding lessons."

These weren't a bunch of cowgirls heading out to a roping class. The girls that went to Stephens College came from big cities; their folks sent them to college for culture, class *and* an education. For some girls, that meant learning to ride English or fancy style dressage.

Luther continues, "One time, I was riding with him on the

bus and he decided to take a different route out to the horse barn. I have no idea why, but he went over by the Christian Female College; today this school is known as Columbia College. This was way off of his normal route. He tried to drive through a big stone archway back out onto the main road. He got that damn bus full of girls stuck in the archway! Both schools had to come and help get the Stephens bus out of the Christian College archway. It was a real fiasco and L— didn't even get fired over the ordeal. Today, he probably would have gone to jail for kidnapping all them young college girls! Hell, I guess I would have been his accomplice! All in all, he was a good friend of mine, and even though he was a married man, we had a real good time running around together. I sure learned a thing or two about getting out of trouble from my buddy. Nobody was as good as him at getting out of a tight spot."

CHAPTER 28

HAPPY 25ᵀᴴ BIRTHDAY

Joan's fat hamburgers are ready, and she sets a plate out for each of us. Fat hamburgers are another one of her unique meals. Because she cooks them slowly and doesn't drain any fat away, the burgers are almost deep-fried. I put ketchup and cheese on my burger. Luther is more complex; he adds onions, pickles, ketchup and cheese.

While stacking condiments, Luther says, "I went along like this, raising hell and going on lots of dates for three years. I had no intention of settling down. I thought playing cards, working at the sale barn and running around with guys like L— in fancy cars was pretty fun. But then, I turned twenty-five. To me, it was just another birthday, but my mama saw things differently.

"'Luther, you are twenty-five years old now,' she told me. 'It is time to start looking for a wife.'

"Then, she kicked me out of the house. It was 1959 and I'll never forget it. I wasn't even living at home anymore, but I liked having my bed and a few of my books stored at home. I was actually living in Columbia with a group of guys, but she sure made a statement when she boxed up all my things. To make things worse, she was ready for me to find a wife. At the time, the only prospect I had for a wife was Joan. Things weren't looking too good with her, on account of me missing a few of our dates. Her mama wasn't real fond of me, either.

"It was one of those deals where everyone in Centralia was trying to set me and Joan up. My mama, Esther, and even Joan's hairdresser Wilma. All these old biddies thought me and Joan would make a fine couple. While Joan was finishing high school, we went on a couple dates. I liked her, but I didn't want to get married. Joan's mama had a real strict curfew and sometimes, I didn't get her home on time.

"Me and Joan rode around on nice summer nights with the top down on my car. We'd stop at the convenience store and I would buy Joan little malt liqueurs. They were in little bottles only about two inches tall. The liqueurs were about twelve percent alcohol and pretty strong. She loved them little drinks!"

"Oh, you liked them, too," Joan says defensively. "It wasn't just me!"

After fifty-some years of marriage, their quarrels seem more like flirting.

"Sometimes," Luther says with an ornery grin, "me and Joan went out parkin' up on Highway 22. Our favorite spot was right across the railroad tracks in a field that belonged to Dale Hamilton. He had his registered Angus cattle in that field, but we didn't pay much attention to the cows."

He laughs, and I look at Joan to see her expression. She pretends she doesn't hear and reaches for a jar of pickles.

"One evening, a snowstorm blew in; we should have tried

to leave sooner. By the time we was ready to go, my car was stuck in a drift. I backed up and spun the wheels over and over. I got the car stuck real bad. To make things worse, it was stuck right on top of the railroad tracks. The Moonglow bar was just up the road. About the time we got stuck, my friend Shorty came driving down the road. He tooted his horn at us and went on by. He had no idea we were stuck. I could have used a little help."

"Too bad you didn't have a cell phone, you could have called to ask him for help," I say, although Luther still doesn't regularly carry a cell phone even in 2014. He keeps a phone in his feeding pickup, but it is more for outgoing calls than incoming.

Luther agrees, "That would have been nice. Instead, I had to run home to get a truck to pull us of the snowdrift on the tracks. I only lived about a mile away, so I jumped out and ran home in the snowstorm. I made it back with the truck and I pulled the car out. After I pulled the car out, I drove the truck home and she drove my car. When I got back into the car, Joan was mad at me for leaving her all alone in a snowstorm. She said, 'What would I have done if a train came?' I wasn't feeling real sympathetic after I just ran a mile through the blizzard, so I said, 'Well, I hope you would have enough sense to get out of the damn car!'

"I got the car off the tracks, but we were way past Joan's curfew. She was nineteen years old and her mama still insisted that I have her home by twelve o' clock sharp. I knew we were past curfew, but there was not much I could do. We finally made it to her house and I started walking Joan up the porch steps to the front door. About the time my foot hit the second step, Joan's mama was already out of the house. She slammed the door behind her. Even in the dark, I could tell she was mad as a hornet. She beat me to the top step, so she had a height advantage. Before I knew what happened, she grabbed ahold of my ear and gave it a sharp twist. 'Luther, it is one o'clock in the morning!' She demanded, 'I don't want you bringing Joan home that late ever again. You hear me?'

"Here I was, twenty-some years old, getting my ear yanked on like a grade school boy. Joan's mama was barely five and a half feet tall and she weighed less than 120 pounds. But based on my predicament, I had to agree with her.

"I said, ' Yes ma'am,' as quick as I could.

"Boy, was my ear smarting! When I got back over to Columbia, I had another good story for the boys. They just couldn't believe that I got run off the porch by Joan's petite mama. Even after our missed curfew, all the old ladies in town were still scheming for me and Joan to get married. I tried my best to ignore all their fussing. When Joan finished high school, she decided to go to airline hostess school in Kansas City."

Joan liked Luther at the time, but she figured if he was going to drag his feet – she was gonna keep on living her life. She wasn't going to be waiting around town for Luther to grow up and be ready to get married! Being an airline stewardess sounded like a fine adventure. Joan had quite a bit of wanderlust. She had already switched high schools just to have a chance to visit Washington, D.C. Her move to Kansas City for airline school was small in comParison.

Luther says, "Back then, to become an airline stewardess, the girls could only weigh 120 pounds. So, Joan had to lose a little weight. She went off to school for a few months and came back a stewardess. She wore a fine looking suit and skirt."

From his seat at the kitchen table, Luther points to a photograph of Joan in her stewardess uniform. The black and white photograph hangs on the wall in an orange quilted frame.

"She told me that Frontier Airlines hired her to work from Phoenix, Arizona," Luther continues. "Even though I didn't want to settle down and get married, I was sad to see Joan go so far away. The night before she left, she gave me a gift. She said, 'I don't want you to forget about me.' Her gift was a long body pillow. She used her little green pajamas to "dress" the pillow and she sprayed her perfume on it, too. She teased, 'Now, you don't have to be alone when I'm gone.'"

What is better than leaving behind the pokey old cowboy who doesn't want to settle down? Giving him a reminder of what a cute, good-lookin' gal he is missing out on. I realize most grandkids don't learn these details of their grandparents dating relationship, but I'm sure glad I got to hear about "the little green pajamas." There's a lesson here for every girl who ever found herself in love with a pokey man: if you want to snag a husband, don't wait around for him to make up his mind. Then, give him something to miss while you're gone!

Joan wearing her airline stewardess uniform.

Luther arrives in Phoenix in 1959.

CHAPTER 29

A ROAD TRIP

Luther says, "After she was gone to Phoenix for a few weeks, I started missing Joan. I called her up and made plans to visit for the week right after school got out."

I don't interrupt Luther while he's talking, but silently I cheer for Joan. I think to myself, "Look at you, Joan, your plan is working. It sure didn't take long to turn this stubborn cowboy into a love-struck sweetheart. A few weeks ago, he wasn't ready to get married. Now, you've got him driving halfway across the United States just for a visit."

"I started driving toward Phoenix to see her," Luther says. "On my way, I stopped in Dalhart, Texas. I found a motel right across the road from the livestock auction. The next morning, I spent a few hours watching the cows sell before I headed down the road again."

Joan interrupts him, "Sierra, did you ever know your

grandpa has a gift for finding livestock auctions? It is the darndest thing. He can pull into a town that he's never seen before in his life and drive straight to the sale barn. He won't even make a wrong turn. It's like he can smell the cattle or something. We used to go on family vacations and we all got so mad at him. We wanted to stop a museum or something. Instead, he drove us right up to a stinky ol' sale barn."

"I am pretty good at finding sale barns," Luther admits. "But when I got to Joan's house, the first thing I noticed was how pretty she looked. I wish you coulda seen her! She was just a little ol' thing – she had the cutest hips and a tiny waist. She had gotten tan from the Arizona sun and her legs were real dark. Whoooooo-weee! Boy, was I glad to see her! Then, just a few minutes after I arrived, I met another man. He was at the house Joan shared with her roommate. He made me nervous. Did I just drive across the country to visit a girl who had a new boyfriend? Had I already missed out on my chance?"

Again, I find myself cheering for my grandmother. She'd made Luther jealous and she didn't have to say a word. She listens to Luther tell the stories of their courtship without comment. If she wants to add anything, she hasn't interjected so far.

"Joan had lots of fun things planned for us to do while I visited," he says. "She took off a few days of work. We went and ate at some place called the Red Apple; you could toss your peanut shells on the floor just like at Texas Roadhouse. While I was there, we kept talking about getting married. It was the same stuff I'd been hearing from my mama and the old biddies back home."

Luther's parents eloped to get married at age eighteen; Luther was already twenty-five. By this age, L.W. and Louise had several children.

Luther says, "I looked up the Trevar Sale Barn during my visit. It was kind of a famous place and I wanted to go. Like usual, it was pretty easy for me to find it. When the subject

of marriage came up between me and Joan, I told her I would
go down to the sale and talk it over with the cows. They
had sales there twice a week. When I got back from the cow
sale, half a day had passed. She was a little upset with me
for spending too much time at the livestock auction. I guess
I was kinda rude; going down to the sale barn after Joan had
taken time off work to be with me. Then we talked about
getting married again.

"I said to Joan, 'The cows said it is not time to get married.'"

I want to tell my grandpa, "Come on you big chicken – buy
a ring already! Stop dragging your cowboy boots in the dust!"
But I keep my thoughts to myself.

"After our time together, I went home and started working
again," Luther says. "But, before long, I started to miss Joan."

He was also worried about the other man who had been

183

visiting Joan and her roommate.

"One day in July," Luther says, "I finally got the nerve up to call Joan. When she answered, I said, 'If you'll come on home, we will get married.'"

I think to myself, "Finally, that took forever!" I'm proud of Joan for her big move to Phoenix. I treasure the photos of her dolled up in her airline stewardess uniform. She carries a special dignity in those photos. She holds her head high, proudly knowing she's not waiting around on some college boy to grow up. Because their dating and proposal took place during the 1950s, I'm sure most folks expected Joan to wait around Centralia while Luther decided to propose. But she had other plans.

Luther says, "After we talked, Joan sold her car, quit her job and came on home to Missouri. She was home before the end of the month. Sometime after she arrived in Centralia, I finally asked her about the other guy back in Phoenix. He turned out to be her roommate's cousin. Or, was he gay? Hell, I can't even remember which. It does not matter; the point is, I was all worked up about him being sweet on Joan for nothing."

Joan and I smile quietly at one another, not as grandmother and granddaughter – but equals. For a moment, we are two women laughing secretly at Luther's petty jealousy. Normally, I would help her clear the dishes, but since I'm taking notes, Joan picks up the table and cleans up her kitchen.

Luther says, "Me and Joan told everyone in town that we was getting married, but I knew we couldn't get married in Centralia. I had done too many ornery things to all my friends! Once, during a friend's ceremony, I went out to their getaway car and put a little dead pig under the luggage in the trunk of his car. It took him a few days to figure out that it was there. By that time their luggage, trunk and the whole car was getting pretty rank. On account of incidents like that, I decided it would be safer to get married outside of town. I was afraid of what they might do to me on our wedding day!"

Joan, fourth from right, at her graduation from airline stewardess school in Kansas City, Missouri.

Luther recalls Joan was especially tan and beautiful when he visited her in Phoenix.

Although Joan didn't get to have a traditional wedding, she did receive a set of wedding dishes. This pattern, called "Franciscan Desert Rose," is still fairly easy to find at antique shops. In 2014, a twenty-piece dinnerware set with service for four might sell for $200, a set of salt and pepper shakers for $20 each and a butter dish for $50.

CHAPTER 30

LOOKIN' FOR A PREACHER

Luther and I leave our seats at the kitchen table and move to more comfortable chairs in the living room. As usual, Joan joins us after the kitchen is tidy.

"We planned to find a church on the way down to Georgia," Luther says. "We would get married on the way and then drive south to meet the rest of Joan's family. Before we left, we went over to Columbia and got our marriage license. It only cost us $3.25. But, I guess it don't mean much that ours was just a couple bucks, because we don't even have it anymore. In all of our life, Joan has kept just about everything she ever owned. But we cannot find that damn certificate. We haven't seen that thing in over forty years."

Joan explains that the last time she saw the certificate, it was in the top drawer of a dresser in the home they lived in previously.

Luther says, "We got things lined up for our trip and planned to leave on a Saturday morning. My sister Rosemary tells everyone that before I left to pick up Joan, I changed my clothes five times. I was a nervous wreck. I picked Joan up and then we stopped at my mother's house to say goodbye. All the aunts and my mama were fussing over the two of us."

Luther's mother straightened his jacket, and acted as if she was going to lick her finger and wipe a smudge of dirt off his face. He resisted, embarrassed, and all the women laughed. It was a smug laugh paired with knowing smiles that said, "Don't you forget, Luther, we set this up!"

Luther says, "Those old ladies were pretty proud of the match they made. We started driving south and I learned pretty quickly it is awful hard to find a preacher to marry you on a Saturday morning. I was nervous about getting married, and knocking on the door of a church looking for a preacher every half hour was about to do me in. Then I learned it was pretty hard to find a preacher to marry you on a Saturday afternoon, too! We had been driving south through Missouri for more than four hours with no luck.

"'Joan,' I said, 'It is getting late. How about we just keep driving and find someone to marry us tomorrow?' I found out that was the wrong thing to say. By the look on her face, I knew were getting married on Saturday and not on Sunday – even if we knocked on churches all the way through Kentucky! One more day wasn't going work. We made it down to Cuba, Missouri, and I still hadn't found anyone to marry us. We found another little church, so I pulled over and ran up to the house next to the church. I banged on the door for a long time. Finally, the preacher came to the door in his underwear.

"'Hello,' I introduced myself to the half-dressed man. 'Can you marry me and my girlfriend?' As I said this, I pointed out to our little car at Joan.

"'Sure,' he said. 'What day would you like to be married?'

"'Right now!' I said anxiously, knowing my nerves would never last another couple days.

"He finally sensed my urgency. He said, 'We can do that. Let me go get dressed and get my wife. She can play the piano for you.' I went back out to the car and got Joan. Then the four of us went over to the church and the preacher got everything set up. Before he started the ceremony the preacher turned and looked at me very seriously. I had no idea what he was going to say.

"'Now son,' he said with a southern Missouri drawl, while looking down at me sternly. 'I've got a slowwww ten dollar marriage or a faaaaast twenty dollar marriage. Which one would ya like?'

"I was the most uneasy son-of-a-buck you ever saw standing in that church. I didn't hesitate a moment before I answered him.

"'I'll take the fast twenty!' I said.

"I couldn't take another half hour of this waiting around. Within a few minutes, the preacher had us all married up. I was flat exhausted. Joan swears it was five after one when he married us, but I don't see how she can remember that. I was too damn anxious. The preacher man said to make our ceremony official we needed two witnesses. This was a problem for us because we were five hours from home. We didn't know a soul in Cuba, Missouri.

"The preacher said, 'Well, there are two old ladies rocking on their porch on the other side of the street, I'll just go and get them.'

"The old ladies came over and signed our marriage certificate and we were officially done getting married – the date was Saturday, August 29, 1959. We spent the first night in Van Buren, Missouri. Joan always says I was asleep by nine o'clock."

She says, "It's true! You were asleep. I spent the whole night crying. I just didn't think that was the way it ought to be on the wedding night."

Luther says, "I have to admit, I probably did fall asleep as soon as we found the hotel. It had been the longest day of my life. An eighteen-hour day in the back of the sale barn would have been much easier on an old boy."

Luther looks at Joan and says, "Honey, I don't think you cried all night."

She replies from her seat across the room, "How would you know? You were asleep."

CHAPTER 31

YANKEE MEETS THE SOUTH

"When we arrived in Georgia, it did not take me long to realize I was in a different world," Luther begins. "First of all, the dirt is red in Georgia. At that time, there were lots of little ol' dairy cows picketed out in the ditches eating grass. I swear, every cow in Georgia was grazing alongside the road. I've never seen nothing like it. When Joan and I pulled up to the house, it was dark. We'd had a long day of driving and I was plum wore out. Once, during my time in the military, I was sent on a short dispatch through Roberta, Georgia. I never expected to see that town again, but here I was – getting ready to meet my new wife's family."

Joan and Luther walked up the stairs to the big house and the screen door was hooked shut. They knocked and waited for someone to come to the door. Aunt Belle came and unlatched the hook.

"Oh, Lordy, Lordy. It's Mister Jesus," she hollered. "Mister Jesus, Mister Jesus. You are finally here! I am so glad to meet you, Mister Jesus. We have been looking forward to your visit all week!"

Luther says, "I had no idea what she was talking about, but she wrapped her arms around me tightly for a big hug anyway. During this commotion, Joan was trying to say something, but I couldn't hear her. Finally, I heard what Joan was trying to tell Aunt Belle."

"Aunt Belle," Joan said, "it is not Mister Jesus, it is Mister Angell."

Aunt Belle replied, "I just knew yo' name had somethin' to do with heaven. I'm so glad to meet you, Mister Angell."

"After we cleared up the confusion with my name, we sat in the front room and visited with T.O. and Aunt Belle," Luther continues. "They seemed eager to get to know me

T.O. Gasset, his older sister, Emidel Havis, and Aunt Belle were happy to greet the newlyweds in Georgia.

and I enjoyed talking with them. We had plans to go and meet more of Joan's family. T.O. had three younger sisters and we were going to meet two of them, Aunt Rosie and Aunt Gladys. The next day, we pulled down another red dirt driveway to a fine white house south of Roberta. The house belonged to Aunt Rosie and Uncle Charlie. Aunt Gladys' husband died a few years before, so she moved in with her sister.

"Aunt Rosie fixed us lunch while the rest of us visited. When she brought us our plates, I could not have been more surprised. All that she put on my sandwich was few slices of tomato. There wasn't any meat or cheese – just tomato, onion and mayonnaise! In Missouri, we don't eat tomatoes like they do down South. Those folks love their tomatoes.

"Everybody else was really enjoying their tomato sand-wiches. About that time, I started feeling like a damn Yankee. I could see it was a pretty well-known meal in Georgia. I was hesitant at first, but then I joined all them crazy southerners; that was the very first time I had a toma-to sandwich. I was suspicious at first. But after a few bites, I changed my mind. It gets awful hot down in Georgia during the summertime, and this light sandwich hits the spot. Even now, on a hot Missouri summer day, I still love to eat me a tomato sandwich!"

Joan says, "Uncle Charlie was one of those men who always had a lot going on. He had a box factory, bought pulp-wood, ran timber trucks and owned cattle, too. Uncle Charlie was always jumping from one thing to another. One day he might go buy some timber off a little ol' widow lady, then head down to the box factory to check on things. Later in the afternoon, he might go out and check his cattle."

Luther says, "We visited with Joan's family for most of the afternoon. She had a cousin who worked at the local livestock auction, so I took some time to go down and visit another sale barn. I watched them sell all them cows. They sold quite a few one at a time, probably to folks who were going to picket them out in their yards.

"By the end of the big day, I was happy to turn in and get to bed. We were staying with T.O. Joan showed me where the bathroom was and set out towels for me to take a bath. Joan was still visiting with her dad, but I headed to the back of the house to take a bath.

"I heard another car pull up to the house, but it didn't really phase me too much. I got into the tub to relax – what a weekend. A few minutes later, someone walked into the bathroom. I panicked, but the fellow didn't seem to take notice. He just walked in on a buck-naked stranger who was taking a bath. He sat down on the commode next to the tub and pulled out a cigarette. He lit this cigarette and in a slower, deeply southern accent he said, 'I'm Toady, Joan's cousin.'

"Here I was, hiding under the bubbles and he was acting like we were in a coffee shop. From the toilet he began chatting with me, buck-naked, in the tub. He rambled on and on, 'How is the weather up there in a…where is it you're from again? I know Aunt Rosie told me but I forgot. It is been awful warm down here lately. You know, I sure am glad you and Joan got married. Did anyone tell you that yet? We're all real happy for you two.'

"I was stuck in the tub and completely shocked. Do all people in Georgia converse between the tub and the toilet? I stayed as far below the water as I could. I listened to Toady rattle on about the cows, his dogs and other things I can't remember now. What a week I was having. First of all, getting married was just about enough to make me a nervous wreck. Then, there were the cows grazing in the ditches, the red dirt and tomato sandwiches. After all that, Cousin Toady just about put me over the edge.

"The next day, I learned some people around town called that fellow Crazy Toady. Toady's elevator didn't go all the way to the top floor, if you know what I mean. This made me feel a little better, but I still felt embarrassed every time I thought about him talking to me while I was in the tub.

"We had several more nice visits with Joan's father and family before we began heading home again to Missouri. But, no matter how hard I tried to fit in, I still felt – and looked – like a damn Yankee. I was sure glad to cross the Missouri line after we headed home. Now, some people might consider Missouri the South, but let me tell you – it is not. I've been to North and South Carolina and Georgia. Missouri might as well be Minnesota compared to those states in the Deep South. In Missouri, we don't have a drawl, we don't picket our cows out in the ditches, we ain't got grits, and we don't eat half the tomatoes that those Georgia folks do.

"When we got home, a married couple, the first thing we did was rent a furnished house in Columbia. It was just eighty-five dollars a month. We decided to live there because I was still finishing my degree and it was a short drive to the livestock auction. I went to my classes in the morning and then Joan would pick me up at noon. I changed my clothes and went right to work!"

CHAPTER 32
GRITS AND GREEN TOMATOES

Luther says, "Newlyweds often fight about silly things during the first few years of marriage. For some it is about vacations, work or in-laws. For me and my new bride, the issue was food. This problem developed within the first few weeks of our marriage. I learned Joan and I had different opinions on the subject. My first mistake was failing to consider that my bride was raised in Georgia. That meant she learned to cook and eat in the South. I suppose a more observant guy may have considered this while dating or during our recent trip down to ol' Dixie. But, I didn't put it together until the third time a steaming pile of grits was scooped onto my plate. At that moment, it finally hit me. Marry a southern woman and you'll be eating southern food!

"This was a big problem for me. I realized on our trip to Georgia, I'm a Yankee. I don't have anything against the South, but her southern cooking was strange to a plain ol' country boy like myself. For example, those grits – they are as near to nothing as anything I ever tasted. They taste like air. It seems like a waste of good, plain corn to me.

"The other thing she kept feeding me was fried green tomatoes. I ate a few, but I couldn't get used to them. It seems to me like tomatoes ought to be red, not green. The whole time I was chewing them up, I felt like something just wasn't natural about a green tomato. Just like dogs and cats shouldn't be friends, a ripe, eating tomato just shouldn't be green. Like I said, it ain't natural.

"Joan knew I wasn't real impressed with these southern foods. I think they were two of her favorites from down South. There wasn't a whole lot she could do about it though; she was just cooking for me the way she'd been taught.

"There was one plain, Yankee food she did know how to cook and that was a baked potato. Every night when I came home, I didn't know what she might cook, but I did know there would be a baked potato. Monday, Tuesday, Wednesday; every day, she fixed me a baked potato."

Joan interrupts her husband, "The reason I was cooking you a baked potato every day is because I never knew when you'd get home! You were either at the sale barn or loading out hogs. A baked potato is something you can hold over without much trouble."

Luther continues, "We went along like this for about a month. I was working quite a bit over at the sale, hauling hogs for Central Hog Buyers and trying to finish up my degree at the University. At night, I was pretty wore out from a long day of work and school. One evening, I was very tired. We sat down for supper and she put my regular baked potato on my plate. Then, she took the lid off of a pot and scooped some asparagus onto my plate. 'What in the hell is that?' I said, pointing to the asparagus. I was tired and cross, which is never a good combination.

197

"That comment pushed my new wife too far. She didn't offer to answer my question, but she moved quickly back across the kitchen toward me without one word. We might not have been married long, but I knew an angry woman when I saw one – I messed up. It reminded me of the night her mother grabbed my ear on the front porch.

"Joan reached my side of the table and took my plate, snatching it out from in front of me. Before I knew what was happening, the back door was shoved open and out the door it went. My baked potato and asparagus went sailing. She flung the food from the plate out into our backyard.

"During the month we'd been married, I thought the only mistake I'd made was marrying a southern cook. I sat there at the table stunned. As my food hit the grass, I realized my second mistake: this ol' Yankee was going to have to be more polite to his southern cook or else I might starve. I'd just worked all day long and I was mighty hungry. Back then, I was six foot tall and weighed only 135 pounds. There I was, a skinny little fellow with no supper. I knew if our marriage and my lanky frame were going to survive, we had to get this worked out. I wasn't going to last long on fried green tomatoes. It would be even worse if my food ended up in the yard twice a week.

"Alone in the kitchen, I grabbed a blank sheet of paper. I sat back down and wrote out a list of thirty days' worth of meals. It was all plain, Missouri food. For the next few months, I instructed my southern bride to cook for us right off the calendar and nothing else.

"In return for taking the list, she instructed me on how to talk to a woman about her food. No more cursing or criticizing. We still have that meal calendar on our fridge after all these years, but we don't need it these days. It might surprise you, but I no longer weigh 135 pounds!"

He rubs his huge, round stomach and laughs.

Luther says, "Eventually, I did come around to some of her southern cooking. Now, one of my favorite meals is her southern-style shrimp Creole. It is full of vegetables, shrimp and kielbasa sausage all served over white rice. And, I do love me a tomato sandwich on a hot day. I think I have sure come a long way from my simple Yankee roots. I even learned to eat broccoli and cauliflower with cheese sauce, but…to this day, I still don't like grits and I won't eat asparagus!"

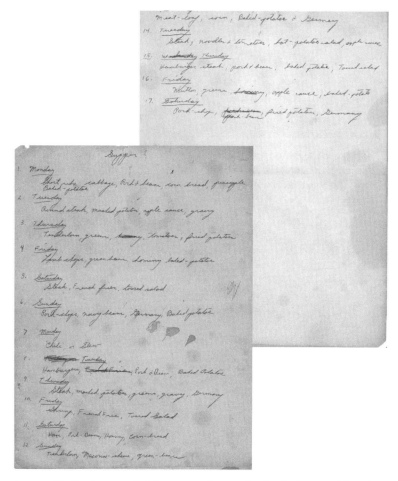

Joan and Luther's menu. Neither remembers eating lamb chops – although it is on the menu.

After living in the rented apartment in Columbia, Joan and Luther moved to a little farm house on "The North Farm" located north of Centralia on Highway C. After she finished decorating their home, Joan took these special photos. (Below) Luther shows off a new bruise from a "cow wreck" at the sale barn.

Joan recalls, "The house had running water, but when the cattle would all come up to drink at one time the well would go down. This caused the pump to quit. Then, I had to go outside to prime the damn pump to get the water going into the house again."

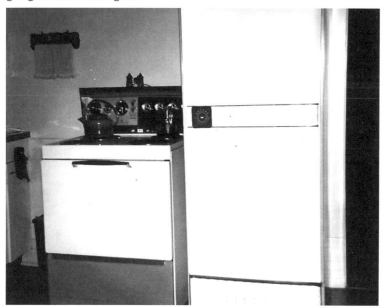

To accommodate the small kitchen space in the farmhouse, Joan moved the kitchen table out onto the porch that was located on the north side of the house. She sewed the curtains for the porch to make it more welcoming.

The quilt covering the bed was made for Luther by his mother, Louise "Honey" Angell. The colors of the quilt were shades of green, brown and white.

WATERING HOLES

Photo by Noltey Hawkins.

CHAPTER 33

THE CORONADO INN & JACK'S GOURMET RESTAURANT

Buddy says, "You might notice that Luther has a lot more stories about going out, shooting craps or stopping at Jack's than I do. That's because he had a lot more opportunities. I was always in Centralia at the Hog Buying Station on Tuesdays and Luther was at the sale. Then, on Wednesdays he finished up in Columbia after the sales and I went back to Centralia to finish up the books on that day's hogs. On the rare occasion that we rode to Columbia together, Luther would say to me, 'Buddy, let's stop at Jack's and have a beer.' Well, everybody knows that Luther Angell has never had just one beer in his entire life! He'd get to having a good time and I was ready to go home. We had lots of work to do the next day. About the fourth or fifth time I got stuck at Jack's all night with Luther, I started driving separately over to Columbia."

With his brother now driving himself, Luther was free to stay out in Columbia as long as he pleased. At the end of a long day at the Columbia Livestock Auction, Luther grew fond of saying, "Well, I'm going now, I guess I'm gonna stop by the hayfield on the way home."

The "hayfield" is just a clever code name for stopping at the bar to have a drink and visit for a while. Through his seventies, Luther's hayfield of choice has been the Centralia Country Club. It is quiet and just a few regulars stop in for a drink. Most of them are cattlemen, farmers or Centralia businessmen. However, over his lifetime, Luther has enjoyed many watering holes in addition to the Centralia Country Club.

"It seems like there has been a beer joint where The Coronado Inn stood in Columbia, Missouri forever. Longer than I can remember for sure and as long as anyone I knew could remember, too. That makes for a long time!

"In the early 1950s, it was a popular 3.2 beer joint for the University of Missouri college students. At that time, all that a beer joint was legally able to sell was 3.2 percent beers. It wasn't until in the mid-'60s that the City of Columbia approved selling liquor by the drink."

"What do you mean 'liquor by the drink'?" I ask.

"Well, before that if you wanted to have a whiskey and Coke, you couldn't order one. That's liquor by the drink. All the bars could sell was beer with 3.2 percent alcohol. Folks got around it though; we would buy our liquor at the filling station down the hill and then order a soda. We mixed our drinks under the bar."

"I didn't know there was a time when bars could only sell beer," I say.

"Yep, I have a lot of good memories there but some of the

best were in the late 1960s and early 1970s. In 1968, a fella named Jack Crouch moved into Columbia from California and bought the popular local honky-tonk. However, Jack's expertise was in running restaurants and he had plans of adding on and upgrading the Coronado to the first gourmet dining establishment in Columbia. To Jack, his fancy restaurant just happened to have a bar; this place was no longer just gonna be an old honky-tonk out on Old Highway 40. The bar kept the name of The Coronado and they added a restaurant called Jack's Gourmet Restaurant.

"Jack was a great businessman and he made it work. It was a white tablecloth kind of place, with candles and polished silverware. The women who dined there were elegant; they wore fancy dresses and expensive jewelry. At that time, everyone in Columbia knew Jack's Coronado was the finest restaurant in town. The restaurant had a big wooden dance floor that Jack kept carefully polished. Two or three nights a week, a band was invited to play at Jack's. As the evening meal progressed, they would play softly in the background and some folks would even slow dance as they waited for

Photo by Noltey Hawkins.

209

their meals to be served. This place was now upscale! The beer joint days were long gone!"

Luther still seems amazed at the transformation Jack pulled off, converting the bar into the best restaurant in town.

He grins mischievously, and says, "Now, not far up the road from all the fine dining going on at Jack's was the Columbia Livestock Auction. Like Jack's, the Columbia Livestock Auction was also in a boom-time. The cattle business was good, the sales were big and everybody was making money."

Even though the livestock auction is now closed, it is still surprising to drive down Old Highway 40 and see just how close the CLA and Jack's sit.

Luther says, "However, the people at the Columbia Livestock Auction were quite different than the new crowd dining at Jack's Coronado. But the fancy uptown crowd didn't bother me. Cowboys and sale barn folks are a loyal bunch! And the whole lot of us didn't hold it against Jack for going upscale. Let's just say, as far as we were concerned, Jack benefitted from his new, classy customers in the dining room *and* the old crowd that The Coronado Inn always attracted."

Luther is smiling wide. He has always loved to play the role of country bumpkin amongst his city friends. To me, it seems, it has always been an act – a joke he played on those high-society folks. He enjoyed seeing their noses turn up and faces wrinkle when he spoke of pickin' wool off dead sheep or feeding restaurant slop to the hogs. Jack's new restaurant provided the perfect playground for one of Luther favorite pastimes: teasing city folks.

Luther says, "The 'boys' that work in the back of sale barns work hard…and they play even harder. After twelve or eighteen hours on concrete, in the heat or cold, the boys are not ready for bed. They are ready for a drink! And the nearest place in town that could serve them one was just a mile up the road at Jack's.

"When the sale got over, cattle buyers, a few sellers and the 'boys' would crowd into the bar just like we always had for years. This made all the barmaids really happy. They knew, when the sale barn crowd arrived they would make good tips!"

Luther is proud that he and the sale barn crowd still held the esteem of the staff at Jack's. They might stink a little more than the other folks, but they tipped just fine!

Luther says, "One night, me and the wife was in there eating supper when Jack came over and sat down at the table with us. He said, 'Luther, last week after you boys were here, I did something that I have never done before. Heck, it was something I never dreamed I'd have to do.'

"I said to Jack, 'What did you do?'

"'After you boys left,' Jack explained, 'I had to clean that dance floor with a scoop shovel because there was so much cow shit on it!'

"As best that I can remember, Jack sold his restaurant to some other local Columbia business people, who kept the name and everything the same. He moved down to Joplin, Missouri and opened another Jack's. Years later, he came back to the area and opened a third one in another part of Columbia, out on Highway 63. During that time, Columbia had two restaurants that were named Jack's, which made for some confusion. Jack's new place never did catch on like the first one. The original is still there today, just as fancy as ever, but no way it could be as fun as it once was!"

Luther says, "When the first Jack's was really going big, you needed to make reservations to eat there on a Friday or Saturday night. For some weekends like homecoming, reservations were made weeks in advance. Me and the wife never did make a reservation when Jack owned it. He always found a place for us, though on one condition. "He'd say, 'Now, darn it, if you want to eat…go ahead and eat! But then, get out! I don't want you sitting around drinking coffee all night.'

"I felt pretty lucky that he always squeezed us in, but my wife loves to drink her coffee real slow and visit after we finish dinner. However, Jack was a businessman and I respected that. He wanted to turn the tables at least three times every night. Over the years, I had become a frequent customer and friend, so getting a table without a reservation wasn't a problem, but we weren't going to disrupt his business, at least not by squatting at a table all night.

"I was actually lucky to be getting a table at all after the stunts I had pulled on Jack over the years. Let me tell you about my friend Bo Foster. Me and Bo always seemed to end up at Jack's after the sales. Bo Foster was a regular buyer. He and his dad, Webb Foster, fed a bunch of cattle back then near Auxvasse. I guess we liked it so well partly because the barmaids weren't wearing too much for clothes. They were always the type of girls a fella had no trouble looking at. Now, Bo always talked with a brogueish accent and a small speech impediment. The combination made him hard to understand sometimes. That lisp didn't matter much, cause ol' Bo sure had a way with words.

"One night we were up there pretty late, when Bo looked at one of the barmaids and said, 'Honey, I'd give you a hundred dollars just to see you wiff your clothfs off.'

"Now, she probably should have smacked him just for thinkin' that, but when he said it right out loud, the barmaid just laughed. She didn't realize Bo was probably serious. Bo had another saying that I'll never forget. He told me this just a few times, but there's a lot of wisdom in his words.

"'Luther, every man has got a weakness,' he said. 'Some men for booze. Some men for women. And, some men for gambling.'

"As he talked, I shook my head agreeing with my friend, but he wasn't finished yet.

"'You know somfan,' Bo said, 'I gots all ffreee of 'em!'"

Luther and I laugh at Bo's insight! I say, "Bo is pretty wise to look at himself and see that he doesn't have one weakness – he's got three! What else fun happened at Jack's?"

"Bob Johnson was another one of my good friends. There were a few things that Bob enjoyed: dancing and smoking cigars. He was always a fun fellow to be with, but especially on the occasions where dancin' and cigars were involved. One time, me and Bob stopped by Jack's after a sale during the wintertime. We decided to play a joke on the barmaids. We were in a hurry to get to the dance floor, so we tossed our coats onto the backs of two chairs.

"Every time the barmaids came around to take our drink orders, Bob and I would point to the coats on the chairs, 'Another round for our friends, they are out dancing.' We bought ourselves drinks that way all night long. I'm not sure if the girls ever know that 'our friends' weren't out dancing. Hell, our friends were just coats!

"Like I said, Bob really liked to have a cigar. And, another night when we were in Jack's, he ran out.

"'Damn,' Bob said. 'I'd give ten dollars for a cigar right now!'

"At that time, a cigar would have cost him about fifty cents or maybe a dollar. A few minutes later, me and Bob were still sitting at the bar. One of the waitresses walked up to Bob and she had a lone cigar balanced on her tray above her shoulder. She held the tray lower, so Bob could see the cigar.

Photo by Noltey Hawkins.

213

"He said, 'Now where in the world did you get that?'

"It wasn't customary for bars to stock cigars, even a great place like Jack's.

"The barmaid said, 'I heard you hollerin' about wanting a cigar and what one would be worth to you right about now. So, I went into the restaurant and found an old boy with a cigar in his pocket. I paid him five dollars for it. Now, I am selling it to you for ten. He made five and I made five.' Bob's eyes were about as big as his whiskey glass. He opened up his wallet and pulled out a ten-dollar bill!"

Luther laughs at the memory of his friend paying ten dollars for one cigar; he could normally buy ten or twenty with that much money.

"That gal was a natural capitalist. There's no telling how far she went after she graduated from the University. She probably paid her whole way through college with tips and ten-dollar cigars she sold at that bar!"

CHAPTER 34

THE COVE

"One night after the sale, me and the boys heard tell about the grand opening of a new bar in Columbia," Luther says. "We always enjoyed visiting a new drinking hole, so we decided to head on over after all the work was done at the sale barn. This new place was called The Cove.

"When we pulled up into the parking lot, I saw that only a few of the spots were taken. We walked in and noticed it hadn't drawn much of a crowd. The place was looking pretty dead so far, but the sale was over fairly early so we thought we would give the new place the benefit of the doubt and hang out to see if it would come alive later.

"Even though there weren't many people there, it seemed like the owners were prepared for a large crowd. They had six or eight pretty ladies waiting around to serve drinks. These were real pretty girls, so me and my buddies were glad to find the bar mostly belonged to us.

"Before long, the girls knew each of us by name and had our favorite drinks memorized. My friends were really good tippers and we were getting plenty of service. Within a couple hours, my friends and I were having a real good time. The place had started to fill up while we weren't paying real close attention.

"At some point, while I was still not paying attention, a news crew showed up to publicize the opening of Columbia's newest business. By then, all the staff now knew me and my friends and they directed the cameramen right over to our table. Boy, did that get the sale barn crowd excited. Keep in mind now, this was before cable and satellite TV with over two or three hundred channels to choose from.

"Back then, there were about three local affiliate TV stations: ABC, NBC and CBS. We were gonna be on TV on the late night local news, which was a big deal. I'm not much for being the shy type. As the camera was panning around, I grabbed one of the barmaids around the waist and pulled her over to me. She sat on my lap, taking down drinks on her notepad, not missing a beat and playing along while my friends and I ordered another round. It was all in good fun and the cameramen got their clip for the local news story.

"That night, on the ten o'clock news, the local channel did a story about The Cove's grand opening. The reporter explained the location of the new business and some of the featured specialties while they showed a shot of the parking lot and the outside of the business. Then, the owner invited everyone to come on out and have a good time at the new bar. While he was talking, they started showing the film of some of the fine fun-loving folks who came out for the grand opening. The TV set had my happy face plastered all over the screen with that pretty waitress right there on my lap. I don't think everyone in Columbia and Centralia was watching the same ten o'clock news channel, but it sure seemed like it later. I think more people saw my goofy face on the TV that night than I normally see in a whole year.

"For whatever reason, my wife did not watch the news that night. You would think that would be a good thing for me, wouldn't you? I think it would have been better if my wife saw the described film footage herself. There is no way it could have been any worse than the multiple phone calls she started to get at the house right after the evening news. Most of the phone calls went something like this:

"'Joan,' they'd say, 'Where's Luther at tonight?'

"She told them innocently, 'He's over at the sale workin' late.'

"'He may be working, but he sure ain't at the sale, honey. We just saw him on the ten o'clock news, and that gal sitting in his lap didn't look like you!'

"This was one of those times that a fella might think that he has too many 'friends'. These folks were just too eager to report on the nightly news.

"Somehow, I think the way my 'friends' were describing what they saw on the news was much different and terribly exaggerated. One thing I know ain't exaggerated, Joan wasn't very happy with me when I got home that night. Thanks to all my friends!"

Luther has never been shy in public. He and Joan call their public displays of affection, "Playin' kissy face." It still happens today from time to time!

CHAPTER 35

THE RED ROOSTER

"The Red Rooster was run by Doc Fenton's uncle. Doc Fenton was the veterinarian for many years at the Columbia Livestock Auction. So, when we was at the Red Rooster, it felt more like a family deal than just another bar.

"Now, all my memories at the Red Rooster involve my friend Woody. He was quite a fellow. When he'd get to drinking pretty hard, he liked to preach funerals. Now don't that sound depressing? Who wants to sit at a bar and listen to a funeral. Well, Woody's sermons were funny. He'd pick one of our friends and start preaching like it was their time to be up in heaven.

"He'd get up on a chair and say, 'Quiet down, quiet down. My friend Luther has gone up to heaven today. So, here we are to celebrate the life of our friend. Now, Luther was known for....'

"Woody would just ramble on like that and it was real funny. But one night, he started giving a funeral and the Red Rooster was hosting a private party in the next room over. They kept telling ol' Woody that he couldn't be preaching funerals because it was too loud.

"About the third time they asked Woody to be quiet, he got a little mad. He hopped down off his chair, walked over to the bar and said, 'I ain't gonna be quiet. Just what exactly would you take for this damn place – I'll just buy it!'

"The thing was, Woody was serious. He would have bought that bar right then if the offer was made. Now, there was one other time when Woody had way too much to drink. This wasn't the kind of night when Woody got up to preach a funeral. This time, me and David Veemer decided it was time to take Woody home. Woody kept telling us he was not ready to go, but we loaded him up anyway. He lived pretty close to the bar, so we just ran him home, put him to bed and then headed back. Me and Veemer were still having fun, so we headed back to the Red Rooster. We didn't make any stops. We just drove right back. Now, I'll never know exactly how Woody pulled this off, but by the time we got back out on the highway – there went Woody. He crossed the damn median in his truck. I mean he went right across I-70. He beat us back to the bar! I guess he wasn't ready to quit."

"That's crazy," I say. "I-70 is such a busy road. And where were the police? Today, he would have ended up in jail for a decade after a stunt like that!"

As usual, Luther reminds me, "It was a different time. Did Mamoo tell you about the time I took Justin's birthday cake to the Red Rooster?"

I nod my head yes and tell him I don't particularly like this story.

He says, "Well one time, it was Justin's birthday. So Joan had a fancy cake baked over in Columbia. She sent me to get the cake on a Saturday afternoon. After I picked it up, I thought I'd swing by the Red Rooster for a drink. By the time

I finally got home, Justin's birthday party was already over. I brought home a half-eaten cake and a melted pail of ice cream. I'd been over in Columbia eating cake with the barmaids."

There were times when Luther's stops at the watering hole got in the way of family time. This story has lived through the decades as a reminder of when things go a little too far. I can't tell if Luther regrets this decision, he doesn't say, but I'm guessing that by retelling the story, he is subtly admitting that this was a bad day. In my own life, this event was a bit traumatic for my father. He's never forgotten the cake and ice cream, and he's never missed a birthday party for any of his four daughters. Usually, he's there at the center of the party, cutting the cake and passing it out to the friends and the birthday girl.

Luther says, "There was one other time I got in trouble because of a cake at the Red Rooster. One of the barmaids figured out it was my birthday and she baked me a cake. While I was there, we all ate about half of it and I took the rest home. When I walked in the door, my wife was not happy to see me with a cake under my arm. She said, 'I wanna know how come that waitress knows you so god damn good that she's baking you a cake.' Now, today, I can't even remember that barmaid's name, but I do remember the cake. It was an applesauce cake and it was good."

I'm glad when Luther stops telling stories about getting into trouble and hurting people's feelings. I'd rather hear the funny ones. These stories show the downside of the good times at the watering holes.

Charles "Buddy" Angell and his family. Left to right: Sherry, Buddy, Scott and Lori, circa 1975.

CHAPTER 36

THE CENTRALIA
COUNTRY CLUB

Luther says, "The country club used to be a pretty fun place to get a drink on the weekend. I was a regular there and one of the waitresses there was named Star. She was a college girl from over in Columbia. Boy, did she know how to work a crowd. There's no telling how much tip money she would make in a weekend!

"At that time, me and Buddy were still partners on the western store. High-waisted, tight black leather pants were really in style. They sold for about $175; they probably go for more like $300 now. One night, I was up at the country club – probably half soused – and I decided Star needed a pair of them leather pants. If she was making good tips in a pair of blue jeans, imagine what she'd make in a pair of them fine-looking leather pants! I drove up to the store with Star so she could get a pair. She took forever putting them on.

"When she finally came out I asked her, 'What took so long?'

"'Well,' she said, 'they were so tight, I had to take my panties off to get them on.'

"I'll never forget her saying that – boy, oh boy! Now, don't you go thinking I'm a creepy old man, giving a college girl new pants! I had other plans in mind. I thought this looked like a fine opportunity to pull a joke on my little brother. Since I gave her the pants free of charge, I asked Star to be sure and thank my brother for the pants the next time she saw him at the club.

"The next weekend, my brother and his wife walked into the country club. Just as usual, Star was there working the crowd and having fun. It looked like my plan would work out just as I hoped, but just to be sure, I went over and reminded Star to thank my brother Buddy. She walked right over, threw her arms around his neck and said sweetly, 'Thank you for the leather pants, Buddy!'

"Buddy had no idea what she was talking about, but I sure put him in an awkward spot. There he sat, red as a beet, wondering what in the hell Star was talking about. Of course, Buddy's wife was none too happy with her husband. What was her husband doing giving leather pants to the barmaid? Poor Buddy had no idea what had happened, but of course it took quite a bit of convincing to get his wife to believe that. I just sat back on my bar stool and laughed!"

CHAPTER 37

THE ROYAL ROOM

"One night," Luther begins, "there were four or five of us over in Mexico, Missouri, socializing at a place called the Royal Room. Our friend Oren was there and he was making payments on a laundry mat in Auxvasse, Missouri. For some reason, Oren just couldn't get the place to make any money."

"Oren said, 'I've made fifty-six of the sixty payments on that place – if someone would take on the last four payments, I'd give them that danged place! I just can't make it work.'"

Luther Angell, Don Dick, Joe Clithero and Kenny Moss were all sitting around the table. The four decided to take on the last few payments and start a new business venture. None of them knew anything about laundry mats, but they had drunk just enough to give the project a shot.

Luther says, "We made the last few payments on the laundry mat, but before long we started having trouble making any money, too. It seemed like the more money that went into the laundry mat, the more money that came out of the laundry mat! Pretty soon, we figured out we was in a losing venture. It was a real mess."

It turned out there were more than a few reasons the laundry mat wasn't profitable. The owners eventually got wind of the fact that their bookkeeper and store manager was paying his personal bills around town in quarters, dimes and nickels. Auxvasse also had a funny rule at that time that didn't allow for a commercial break in the water fees, so every new load of laundry really was a losing venture! The water bill was higher than the quarters somebody paid to do a load of laundry.

Luther says, "One night, we went to refill the change machine and it turned out that we'd been robbed there, too. Somebody figured out that a Mexican penny is the same size as an American half dollar. We opened up our empty change machine and found a pile of pennies instead of the correct amount! I tell you, the more we put into that place the more that came out! It was always something. Another night the cops called me because they had found a bum sleeping in the commercial-size dryer. He figured out that was a pretty warm place to sleep!"

The four partners accepted that their project was a failure, so instead of trying to make money, they went back to doing what they'd been doing so well on that first fateful night in the Royal Room – drinking.

Luther says, "We started going over there once a week to see how much money we'd made. We'd buy a fifth of whiskey and head toward Auxvasse. On the way over, we took bets on how much money would be in the washing machines. Whoever got the closest got to keep the pile of change. All we ever did was lose money, but we sure had fun doing it!"

CHAPTER 38

THE 2100 CLUB

"Alright," Luther says, "I've got one more story for you about the 2100 Club and then that's the last of my watering holes!"

"Okay," I say. "Let's hear it."

"Well," he says, "we'll have to wait until your Uncle Jon is around. This story involves him, too!"

A few weeks later, Jon, Luther, Joan and I are eating some lunch together. I remember that there's a story I'm missing out on because these two needed to be together to tell it.

"Okay, we are all here now," I say. "What's the story on the 2100 Club?"

Jon begins, "Well, in 1987 I was going to school in Springfield, Missouri. I would come back home to Centralia to visit on the weekends. One night, I got pulled over. The patrolman came to my window and clocked me going 82 miles per hour in a 55 zone. You see, it is not that I have a lead foot, but I was going down one of those long, steep hills going into Jefferson City. I knew this was going to be a huge fine and quite a ticket, but the strangest thing happened when I handed the patrolman my license. 'Jon Angell?' he said in a gruff voice. 'You wouldn't happen to be any kin to Luther Angell, whoudja?' When he asked me that, I was pretty sure I wasn't going to get a ticket – I was going to jail! I tried to be clever and said, 'Well, sir, does that help me or hurt me?' I could tell my attempt at humor annoyed the patrolman. He said, 'Just answer the question, young man.'"

Jon looks at Luther and says, "Do you remember what I told him?"

Luther says, "No, but I remember how it ends."

"Oh come on, you two," I say. "I've already waited three weeks to hear this story—get on with it! What happened?"

Jon is enjoying his moment in the spotlight. Just like his father, he's learned the art of pausing for suspense and drama in a story.

"I told the patrolman, 'For better or worse, I am Luther's youngest son.' Then the strangest thing happened. He handed me back my license and said, 'Take a good look at my nametag. You remember my name and tell Luther he owes me a twelve-pack of beer.' Then the patrolman kinda grinned, 'I better not catch you speeding on my road again. You get your little ass on down the road.' When I got home that weekend, I had quite a story to tell my parents, plus I had a few questions for my Dad. How did he know this guy well enough to get me out of a ticket?"

Luther says, "By the time Jon finished telling me and Joan his story, I knew exactly which patrolman pulled him over. Jon didn't even have to tell me and I knew who I owed a twelve-pack of beer. For the sake of his privacy, I won't tell you the name. But let's just say, over the years I've had many friends in law enforcement. This wasn't the first time I had owed something to that particular lawman. One night after a particularly late sale, I ended up out at the 2100 Club having a couple of drinks with my buddies from the highway patrol. I don't remember exactly the time, but it was well after midnight when I told the boys I was heading home. I had to be at work early the next morning. 'Luther,' the highway patrolmen told me, 'There ain't no way that you have to be at work before we do tomorrow.'

"'Yes,' I assured them, 'I do, and with the shape you fellas are in right now, I'm willing to bet that I'll be up and going hours before you.'

"'A bet?' one said.

"I agreed to a twenty-dollar bet and then I asked, 'How are we going to know who won?'

"All he said was, 'Don't worry about it. I'm going to win and I'll see to it that you know that I won.'

"We had a good laugh together and I headed home feeling confident I'd win. They might not even remember the bet come morning. As I left I hollered back at them, 'You can pay me next week after the sale.'"

"At 4 a.m., I was abruptly wakened by my wife pounding me on the shoulder hollering that the police were outside. When I opened my eyes, I couldn't figure out what was happening. There were red and blue lights flashing and fluttering all over the walls and ceiling of our bedroom. For a moment, I thought I was back in the bar; everything was lit up like a dance floor with a disco ball. When I came to, I realized what Joan was saying, 'Luther, wake up! What is going on out there?'

"Back then, we was living in a house with an unattached garage a few steps from our back door, and a circle drive that went all the way around the little garage. I heard the sound of a vehicle's racing engine. It was going real fast right outside our house and spitting gravel from under the tires. The sirens were so damn loud it sounded like they were in the bedroom with us.

"I finally made it to the window and jerked the blinds to the side so I could see out in the yard. About that time, a highway patrol car came around the corner of my little garage.

"'Wwwwwweeeeeooooooo-weeeeeeooooooooo.' My thoughts were interrupted by the blaring siren, 'Weeeeeeoooooo-weeeeeeooooooo.' The patrol car was going around and around the garage, 'Weeooo-weeeooo-weeeeoooo.'

"I knew then what was happening. My friend had won the bet, but I was still in shock. Only a few hours ago, they were bellied up to the bar at the 2100 Club. I figured there was no way he was in any condition to be driving, and definitely not in uniform. Yet, here he was waking up my wife, my kids and probably the dag-blasted neighbors with the sirens and lights.

"I grabbed my wallet out of my britches and headed for the back door, knowing that was the way to end this ruckus. I went to the back door in my nightshirt and I flipped on the porch light. The patrol car came to a stop and my friend stepped out of his car. He walked up to the door and without a word from either of us, he collected a crisp twenty dollar bill from me. After they left, I tried to quiet down my wife and kids. Everyone was awake after all the excitement. I explained the bet to my wife, but she failed to see the humor. I told that story to Jon after he escaped his ticket back in 1987."

Jon says, "I still remember what I told my Dad that day. I said, 'Dad, I sure am glad you paid him on that bet back then, but I sure think we need to send him that twelve-pack of beer 'cause I travel that stretch of highway way too often not to....'"

Luther and friends enjoy a party in the "The Ol' Blue Room" at the Columbia Livestock Auction. These parties were frequent during the 1970s. Luther was generally reserved, quiet and shy – or something like that!

PRANKS

CHAPTER 39

WAS IT WORTH A QUARTER?

"I have always enjoyed a good prank." Luther says. "Lots of people have played pranks on me, but I think I have played even more on others. If you're gonna try and pull a prank on one of your buddies, there is a certain degree of risk involved. It is a lot like the cattle business. Sometimes things do not always turn out like we expect. One prank may have a fellow rolling in the grass laughing. But another may not go as planned. You might find yourself apologizing, wishing you never had such a dumb idea, and wondering if you'll ever grow up and get any smarter."

Luther says, "One time, I was out at a swimming party at Jack Chance's house. Everyone was having a real good time and we were enjoying a few drinks. One of my friends was having a great time until his wife showed up. She was a schoolteacher. She was mad at him for drinking and she wanted to go on home. She had been at a different party and was all dressed up – she had on a dress, panty hose, high heels and her hair fixed up all nice. My friend got a little frustrated with her, so he looked over at me and said, 'I'll give you a quarter if you will throw my wife in that pool.'

"I thought to myself, 'That's not too much work for a quarter; that don't sound too bad at all!'"

His friend's wife was none too thrilled with the idea of getting thrown in the pool, and she was even less pleased that the whole shenanigan was over a quarter! Luther picked her up around the waist and started walking toward Jack's pool. Naturally, the fully-clothed woman began screaming and kicking in protest. Luther ignored her and continued walking, but he had no intention of throwing her in the pool. He just wanted to tease her, have a little fun and laugh.

However, when the pair got to the edge of the pool, the woman looked down at Luther and made one mistake.

She hissed, "Luther Angell, you don't have the guts to throw me in the pool!"

After she made that comment, Luther didn't hesitate for a moment. He hadn't actually planned to throw his wife's friend in the pool, but once she challenged him – he had no choice. Clearly this woman had no idea who was holding her over the edge of a chilly pool!

Luther says, "So in we went! Hell, I jumped right in with her! Boy, when she came up she was madder than a hornet. Everyone else thought it was real funny, except her. She was mad over her hairdo and her good watch getting wet. I walked over to her husband, sopping wet in my slacks and cowboy boots, and collected my quarter!"

The woman continued to be upset with Luther and left the party in a tizzy. The next week, Luther received a bill for her hair, clothes and watch. To make matters much worse, this woman was a local schoolteacher and she had one of Luther's sons in class. Joan was afraid she'd flunk her son because their dad threw the teacher in a pool!

Luther says, "After I paid for her hairdo and a few other things, she wasn't so mad anymore. Within a few months, we were laughing about it. Over time, it got to be something that was real funny for her to joke about. I don't care what it all cost me though – it was worth a quarter!"

CHAPTER 40

THE LITTLE LAMB AND THE HIPPIE

Luther says, "One of the girls who worked at the Red Rooster told me she had always wanted a baby lamb for a pet. This particular waitress was a hippie. She had long, straight hair and tie-dyed clothing. This was in the early 1970s and this gal was the real deal! She lived in a regular hippie commune during that time, when communes were the most popular. She knew I worked at the sale barn in town and that I would be the best person she knew to help her little dream come true. The rest of her friends were city folks. They didn't know much about animals, much less where to get one. Eventually, a lamb did come to the sale. However, it was not a huggable, lovable baby lamb. It was a dirty, wooly full-grown ewe. The July heat was too much for the old, weak ewe and she happened to die in her pen. Rather than waste the opportunity for a good laugh, I called up my hippie friend."

At the time, Luther was in his early forties. Luther and Joan were married and they already had a couple of half-grown kids. In short, he was past the growin' up stage and there was no excuse for his next actions. He should have known better.

Luther says, "I called up the hippie and said, 'We had a lamb at the sale today. If you'd like to come pick it up, I'll give it to you.' She squealed with delight, 'Oh, thank you, Luther! I'll pick it up after I finish my shift.' Later that evening, she pulled up to the sale barn in a Volkswagen station wagon. I told her to back the car up to the load-out chute. I expected her to come alone, but she brought her hippie boyfriend along, too. She was a dainty, small woman, but he was a big, serious-looking fellow. I could tell by looking at both of them they knew nothing about animals. Some of the guys in the back laughed and watched as I carefully put the dead lamb in the back of their station wagon. I explained seriously, 'Now, the lamb has had a very long day. She is very tired. So when you get home, don't try and move her. Just leave her in the car and let her sleep. In fact, let her sleep until she wakes up.'

"This hippie barmaid was aware of her inexperience and willing to learn. She took my instructions seriously, just like I hoped. It was about mid-July and the weather was hot and humid. The next day, that ol' sun came up and beat down on the hippie station wagon. Anyway, you know what happens to a dead animal on a hot day. The lamb got to swelling up and stinkin' real bad in their car.

"They must have figured out what I did, because during the sale the next week, the girl's big hippie boyfriend showed up at the sale. I did not know she had such a strong-looking boyfriend or I might have thought twice about my prank! But it was too late for second thoughts. He brought friends from the commune and they were all looking for me! The sweet waitress wasn't with him and the friends he brought were really big and rough looking."

Most people remember hippies as young kids who were all about free love and peace. There were quite a few of those types. However, what the history class may have missed when explaining hippies were the Manson Murders. Before Charles Manson, hippies were not scary folks at all, sorta like the waitress. They were harmless. Manson was the leader of a cult of hippie types out in California. He was a real strong leader and plum crazy. Somehow, he brainwashed a few of his followers into killing some people. Maybe it was for some type of sacrifice or something. When the news line hit, "Cult of hippies murder innocent citizens," the view of hippies across the nation changed instantly. From then on, everyone was a little bit scared of hippies.

Luther says, "When three or four of them strong-looking hippies showed up at the sale barn looking for me, word got around to me pretty quick. I thought to myself, 'If those California hippies killed people for no reason, what would these guys try and do to me?' I was regretting my little prank. So, I headed for the only safe place I could think of in the entire sale barn. This secret hiding place also happened to be the hottest spot in the sale barn come mid-July. I spent the afternoon sweating in the hayloft, hiding from hippies."

The hippies kept asking the guys in the back where Luther was. The sale barn boys would say, "Oh…I think he is up front."

When they got up to the front, the ladies in the café said, "You just missed him. He went to the ring."

Luther says, "The good people at the sale barn sent those hippies all over the place looking for me. Finally, they got hot and gave up their search. I stayed up in the hayloft for a good hour after I got word they left. I was sure hot, but I wasn't taking any chances. I was scared of being beat up or maybe even murdered. It is a wonder I didn't die of heat and swell up stinkin' like that poor old ewe!"

CHAPTER 41

THE LITTLE PIG THAT WENT TO JACK'S

Luther says, "One night after the sale, me and a couple of cattle buyers named Bo Foster and Jim Busk decided we wanted to do something different. Something fun! These two knew how to have fun and the three of us together was right dangerous."

This was the same Bo who knew his three weaknesses: women, drinking and gambling. Jim Busk was from Boonville and he worked for Swift as a fat cattle buyer who made it to the Columbia Livestock Auction each week.

Luther says, "Me and these boys "borrowed" a little Hampshire pig from the back of the Columbia Sale Barn. He probably weighed less than forty pounds. We cleaned him up real good! After we scrubbed him clean, we took a red ribbon and tied it in a bow around his neck. That just didn't seem to be enough, so we found a can of old red barn paint and painted the little pig's toes. Now that we had our little pig all cleaned up and looking mighty fancy, we wanted to show him off. So the three of us, plus our fancy little pig, piled in my pickup and headed up the road. Our fancy little pig was dressed for a night on the town. We thought he should visit Jack's white tablecloth restaurant. We looked inside to find the place was pretty full; it was the perfect time. I opened the door while Bo and Jim turned our little black and white pig loose into the restaurant.

"At first, no one noticed the pig. After they set him down, he just walked in slowly. Then a waiter saw him, and he lunged and tried to catch our pretty little pig. That startled the pig. He let out a high-pitched squeal, started in with some loud snorting and ran further into the restaurant trying to get away. I watched as he ran under the table and between a finely dressed man's legs. Within a few moments, the restaurant was in chaos and it wasn't just the pig who was squealing. Men were yelling, women were screaming and the pig ran around, pooping all over Jack's fine dining establishment.

"This was another prank that turned out different than I expected. In fact, it was far more 'successful' than I ever thought possible. I carried through with this bright idea only to think, 'I wish I could get this one back.'"

At the time, there were still lots of people in the Columbia area who owned pigs. Luther had thought going into his plan that he would have what the politicians call "plausible deniability." Jack's had quite a few customers who could have brought that pig to the restaurant. But, what Luther didn't think about was that Jack didn't even consider how many people in the area had easy access to a pig. Jack knew that several folks might have dreamed up a stunt like that but nobody else was crazy enough to follow through on it – except Luther.

"Jack had me on the short list before the dust settled," Luther says. "Of course, he was furious with us. I began to regret my bright idea even more as I thought, 'What if Jack never lets me come back? This is my favorite bar; where will I go instead? I've really done it this time.' Now if I were an unlucky guy, the story would end here with me being kicked out for life. Lucky for me, the story didn't end here. Later in the evening, the local news channel got a tip from someone who was at the restaurant about the big disturbance over at Jack's. They came down to the restaurant and interviewed some of the folks including Jack and a few of the staff about the incident.

"That night, leading up to the ten o' clock news there were several promotional spots saying, 'Stay tuned to learn about the little pig that went to Jack's.' During the news they ran with the story and interviews from Jack's place.

"At the time, much more so than today, the local ten o' clock news was kind of a big deal. There were maybe forty or fifty thousand people living in Columbia, not counting the college kids. Even if only half of them watched the news that night, that was quite a few people who learned about Jack's restaurant.

"After the story aired on the news, people were talking about it," Luther says. "People who had never been to Jack's to eat were now coming for supper. They wanted to see the fancy place that the fancy pig visited. Business was even better than before and Jack wasn't mad at Bo, Jim and me anymore. In fact, he was thrilled. Years later he told me, 'That pig was the best advertisement this place has ever gotten.' As far as I know, the little pig that went to Jack's never returned to the back of the sale barn. That night, he went home to the fancy new Whitegate Apartments with one of the waitresses from Jack's. Not many pigs are so lucky, but even so that fancy pig's luck ain't nothing compared to how lucky I am!"

Luther gives his sons, Jed and Justin, a ride on a dappled gray pony.

CHAPTER 42

THE TRAGEDY OF THE CHRISTMAS PONY

Luther says, "During the winter, people tend to hang around after sales longer than usual because it gets dark so early. One of my most memorable Christmas stories is from a Saturday night years ago at a sale barn in north Missouri. On this particular night, we planned to stay late because the sale was small and we wanted to celebrate Christmastime among good friends. The clerk at this barn was a friend of mine. During the sale that day, he had bought a little pony. My friend was looking for a gift for his children and he thought a pony would make a good Christmas gift.

"The evening was growing late and our group was still having fun hanging around the sale barn. Eventually, my friend decided he ought to go home and hook up his trailer. He wanted to haul his newly purchased Christmas pony home before it got dark. We begged him, 'Just haul the pony home tomorrow. Why don't you stay here with us?' Our friend ignored our pleas and went home to get the trailer.

"'While he is gone,' a second friend said, 'I'll give the pony a shot of a sedative and put him to sleep. Then he won't be able to load the pony. He'll have to stay and have a good time with us.' This old boy happened to be a doctor, but not a veterinary doctor. I mean he was a people doctor. I wish I could, but I can't recall this friend's name. I suppose if I could remember, my lawyer – and my wife – wouldn't let me tell you anyway. We were having such a good time that we thought the doctor's prank sounded like a pretty darn good idea.

"'This will be a good one,' I thought to myself as the Doc pulled out his medicine bag. Keep in mind, this was many Christmases ago, probably in the late '60s. Back then, a country doctor always carried a bag full of bottles, bandages and anything else he might need to have ready. The Doc grabbed a syringe and a sedative from his bag. We pulled on work coats and gloves and our clumsy crew left the warmth of the office and tromped out to the back of the barn.

"Most of the animals had been loaded out by that time, so finding the pony wasn't much of a challenge. Our pony prank was off to a good start, but after this, things got a little rocky. The doctor was used to giving people shots, not ponies. He wasn't exactly sure what amount equaled a pony-sized dosage. So, he made his best guess about how much medicine he would need in his syringe. All he had to do was put the pony to sleep for a couple hours.

"The Doc guessed wrong. He killed the pony deader than hell! That night, I had full confidence in the Doc's ability to estimate. In hindsight, I bet the Doc couldn't have calculated a proper dose of anything, considering his drunken condition. Too many hours of celebrating with his friends led to a few too many cc's of sedative.

"This is a good place to put in a few disclaimers. We all felt bad that the pony was the victim of our poorly executed prank. I realize with all the humane treatment and animal rights advocates around every corner these days, a pony prank like this would have had a much different outcome today. I imagine somebody, or several somebodies, would have ended up in jail.

"Please keep in mind that this prank wasn't pony murder. In legal terms, it was a 'misfortunate accident' at best and 'pony manslaughter' at worst. I find it also important to state that at no time during the making of this story did any animal suffer. I'm confident that the Christmas pony's death was more peaceful than most other ponies. I am also confident that the statutes of limitation have long since kicked in to protect me and my friends from the threat of legal prosecution for our actions against said pony party.

"When our friend came back with his trailer, we fessed up to what happened. The evidence was a cold pony and he was not happy with us. He said, 'Now what am I going to give my kids for Christmas?' After that, he went home. Obviously, he was pretty upset with us. We were upset that we were the bums that had ruined Christmas for a bunch of kids. As terrible as it sounds, we got over our upset fairly quickly.

Buddy gives his daughter Lori (Angell) VanMaanen a ride on a white pony. Thankfully, these ponies were not involved in any tragedies.

247

Not long after he had left, somebody else had another great idea. Knowing how well our first great idea for the night had turned out, we probably should have just let it go. We didn't.

"It was far past midnight when we loaded the Christmas pony into the back of a pickup truck. The lights of the house were off and everyone at our friend's house had gone to bed. We coasted into the driveway planning to head back to the barnyard. Then, someone had one more good idea. The family's four-door sedan was sitting in the driveway.

"'Wouldn't it be funny if we put the pony in the car?' a friend proclaimed.

"'He'll remember this for years,' we said, while propping the pony up on its rump in the backseat of the family car. We took the front hooves and put them over the front seat and crossed them. For the final touch, we set his head up on the front seat. When we finished, the Christmas pony was sitting up, just like a person.

"The next morning, our friend's wife and kids hustled out of the house all dressed to go to church. They found an unexpected passenger waiting for them in the back seat. His wife somehow missed our sale barn style humor. Unfortunately, it had been unseasonably cold that December night. We failed to consider that the pony would freeze in that sitting position, meaning he wouldn't fit out of the door.

"From what I hear, they had to take the car apart, doors and all. Twisting and turning the frozen pony yielded no successes. My friend had no choice but to take the pony apart to finally remove him from the car.

"In many ways, this is a tragic Christmas tale – the pony died, the kids lost their Christmas present, we almost lost a friend and the Doc could have lost his license. Yet, the Christmas season has a way of working out, no matter what chaos may arise. Finally, our friend was still in need of a gift for his family. He had the first truly great idea in this whole story. He went out and bought the best, brand-new car off the dealer's lot for the family Christmas gift that year."

CHAPTER 43

A HOUSEWARMING
PRESENT

"For many, many years I have had the same doctor," Luther says. "I'm just an old country boy, so I don't go to the doctor unless it is pretty serious. Even though I didn't go see him very often at the office, Doc and I were still friends because he lived in Centralia. One year, Doc decided to move closer to work and built a brand new house, right on the edge of Columbia. It was a fine house, with three stories; the windows overlooked the edge of the cliff that it was built on. It was built so that you could walk out of the second story right to the yard.

"The year he built his new home, Doc had a housewarming party at Christmastime. It was going to be a fancy party; most of the guests were other wealthy doctors and nurses from Columbia. However, there were a few of us rednecks from Centralia who also attended. Since it was a housewarming party, I needed a gift. The day of the party Joan and I drove our old pickup out to the Amish community in Clark. I bought an old stinkin' billy goat from one of the families and tied him up in the bed of our pickup truck.

"That night, we went to the Christmas party. The house was wonderfully decorated, full of garland and a grand Christmas tree. The front room of the house had a winding spiral staircase that went from the third level to the main floor. The guests were all wearing suits or fancy cocktail dresses. Doc was a wonderful host. As the night wore on, his party continued with fancy drinks and nice food. While everyone was gathered around the beautifully decorated tree, singing Christmas carols, I snuck out to the truck to bring my housewarming present inside. I took him out of the truck bed and led the old, smelly billy goat to the back of the house. As we stood at the top of the spiral staircase, I untied the piece of twine from his neck. Then I left him at the top of the staircase and ran around to the front of the house and rejoined the rest of the party singing Christmas carols as if nothing had happened.

"Shortly after, I started to hear the noise. At first, faintly over the conversation of the party goers, 'Bah....' bleated from above, unseen. Then again, but this time louder and longer, 'Bah...baaaah....' The singing stopped. Louder again, 'Baaah....' A slow hush came over the crowd, as the attention of the gathering turned fully toward the stairs as the old billy appeared boldly in view above and started down the steps, 'bah, baaaah....' The billy descended the steps one by one; the mouths in the crowd also dropped open one by one in shocked amazement. When he reached the main floor, the goat walked right up to the Christmas tree and started eating pine needles, knocking an ornament off a branch. Then, he lifted up his tail and relieved himself on the floor in the middle of the party. Unsure of where to put my fine housewarming gift, Doc instructed his young son follow the billy goat around for the rest of the night, with a dust pan and broom, cleaning up after him.

"Just like Jack and the little pig, Doc was quick to figure out which one of his guests brought this present! He was quite certain it didn't come from any of his Columbia doctor friends.

"Now, Doc had a good sense of humor and a bit of an ornery side, too. He took the joke well, but he patted me on the back, promising to get his revenge."

CHAPTER 44

DOC'S REVENGE

"A few years later, I had a blood clot in my leg. It was pretty scary, because my own father had a serious stroke. I wasn't in the mood for taking a risk, so we rushed over to the hospital. I don't know how nurses do it, but they convinced me to strip out of my pants and boots, trading them for a flimsy hospital gown. There I lay, naked as a jaybird on the cold hospital bed worrying about dying. Boy was I happy to see Doc walk into my room. I said, 'Glad you are here, I need some fixing.'

"The nurse was tending to my IV and Doc looked serious with his clipboard and white coat. Joan sat anxiously over in a corner. The Doc said, 'Well, Joan, there's nothing I can do for this sorry son of a bitch. We've just got to wait this one out. That blood clot is either going to go to his heart and he'll die or the other direction and he'll live. What do you say we go out and get a steak?'

"Now, me and Joan knew Doc pretty well, but that poor nurse turned white as a ghost. I thought she was going to faint! Doc got her calmed down and explained few more scientific things to me about blood thinner and blood clots—but then, that ol' bastard really did take my wife out for supper! I had to lay there waiting to live or die, while those two went out for ribeyes.

"When they came back, Doc cleared me to go home. The blood clot had dissolved. Doc acted as if nothing serious had happened at all. Here I was, ready to read my final will and testament. Instead of letting me feel sorry for myself, he said, 'That was the best steak I've had in months – and such a pretty lady to join me!' I'll say he got his revenge."

Luther's silly moments in life have far outweighed his serious ones.

CHAPTER 45

A PARTY AT DOC'S

"This reminds me of a time at another party with a different friend of mine. We'll just call him Doc. Now keep in mind, a man like me knows a lot of fellows named Doc in his life-time." Luther says, "Don't go trying to guess which Doc I'm talking about. In central Missouri over the past six decades, there have been a bunch of wonderful veterinarians who carried the title of Doc. This particular Doc had been married three, or maybe five, times. Of all his wives, he liked the first one best. Even after all the years they'd been divorced, once a

LEFT: A friend of Luther's bet him that he wouldn't wear this silly-looking hat to his oldest son Jed's wedding reception. Of course, that just gave Luther one more reason to wear the hat. No one can remember who made the bet or what Luther won as a prize. Joan and Luther call this style of hat a "go-to-hell hat," as in, "I'm going out and having fun in this hat and I don't care what anyone has to say about it!"

year they would meet up somewhere for a weekend together. I don't know when they stopped doing that, but for a while it was the going thing.

"One night, there was a party out at Doc's. The way the house was set up it had a kind of half basement. We was down in there dancing and carrying on. It got to be dark out and the lights from the street were shining in. I guess it was bothering us because somebody had the idea to go outside and shoot them.

"Course, I never did shoot a gun very much, but somebody kinda challenged me. 'Why don't you go outside and shoot the lights out?'

"I walked out into the yard, held the gun up into the air and shot out the lights. I turned around and walked back into the party, as if nothing had happened. My friends were shocked. Now, I didn't realize it at the time, but on the light poles that were going by Doc's house, there was a big cable that went all the way from Columbia to Fulton. Turns out I shot that cable, too! That big cable was for all the phone lines between them two towns. For a couple days, there was no phone service between Columbia and Fulton. You know, that was another one of my good ideas that didn't work so well. That was kinda big trouble for a while and I had to lay low.

"I am the first to admit—sometimes my pranks didn't go as planned and sometimes we took things too far. But other times, we got such a good laugh!"

Although Luther was often pulling pranks on his family and friends, occasionally he let himself be the "butt of the joke." In this case, that involved getting dropped into a water tank in front of a curious crowd.

All of the men in the Angell family are known for their uncanny ability to take a nap anywhere – at any time. They also have a gift for falling asleep in two minutes or less. This is called a "fifteen-minute power nap."

These may be the only three known photos of Luther Angell wearing shorts!

THE GOLDEN YEARS

L.W. and Louise "Honey" Angell and grandchildren in 1966; Scott Angell, in stroller; Todd Boender, standing in grass; Tim Boender and Lori (Angell) VanMaanen, standing on picnic table. In background, several Oldsmobile cars sit. Honey always got the new car and L.W. inherited the old one for chasing cattle out in the pastures.

CHAPTER 46

A SUNDAY PASSING AND A
SLOW PASSING

One Sunday afternoon in 1970, Joan and Luther took their boys to Mexico, Missouri for ice cream. This was a weekly tradition. Jon was about three years old at the time. He was learning to speak and for a few weeks he started calling his grandmother, Louise Angell, and his mother, Joan, by the same name: Mama.

Joan says, "On that day, we were driving over to Mexico and little Jon started saying, 'Mama is dead, Mama is dead.' I tried to explain to him, 'No, I'm not dead, Jon. Mama is right here.' But, he wouldn't listen to me."

Everyone enjoyed their Sunday shake and then the family drove back home to Centralia. Shortly after they arrived at home, the phone rang.

L.W. was calling and over the phone he said, "I can't wake Honey up. I think she might be dead."

Joan says, "Luther has never been one to drive real fast, but that trip across town I don't think our car touched the ground three times."

Joan and Luther left the three boys at home and went over to L.W. and Louise's home to see what was going on. Earlier that afternoon, Louise – who many years before earned the nickname 'Honey' after throwing the honey pot at L.W. – had gone to take a nap. She did not wake up from her nap that afternoon. She died peacefully in her home at the age of fifty-eight.

No one could figure out young Jon's premonition – somehow he knew that his grandmother passed away that Sunday afternoon. When asked about the strange incident as an adult, Jon mainly remembers an overwhelming feeling of grief and sadness that his grandmother was dead. He is quick to mention that he has had a lot of unexplained, overwhelming feelings since, "but none have turned out to be so accurate."

As Jon grew up, he spent lots of happy times with his widowed grandfather, L.W. Angell.

Jon says, "It seemed that he didn't like living alone. He was always glad to invite the grandkids out to supper or to his house to stay the night."

One project that became a year-round excuse for L.W. to collect a few grandkids was loading up the trash and boxes from the western store.

Jon says, "He said he didn't want to pay the city to haul off the cardboard boxes, so on Saturday afternoons L.W. hauled them himself. He parked the old farm pickup in the alley behind the store. We would stack it high with the boxes from the week's clothing deliveries. When we were done loading we hopped in and rode to the farm. Of course this drive was a slow process. There were numerous and frequent stops to reload because a few boxes blew off, if not by the wind of our travel, nearly always when we would meet a big truck on the highway. There would be a woooosh of the wind and then strewed boxes along the road!"

At the farm, L.W. had plenty of things he wanted to burn. He was known for lighting fires to dead trees or stumps in the pastures or to log jams in the creek at a water gap.

Jon says, "We piled the boxes on the stump or stuffed boxes and papers between logs and set it ablaze! Building bonfires is great fun for boys and an excellent recruitment tool for grandpas. However, with L.W., this job was so frequent that it became a repetitive chore. The chore was performed again and again without incident, except for a few times. Upon leaving the blaze to work on the stump, the winds might pick up and blow a burning box across the pasture. More than a few times, L.W.'s pasture or hay field was unintentionally burned black. The dollars and cents of the matter was that L.W. should have paid the garbage truck to haul off the boxes, but then again, this wasn't ever really a chore he did to save money."

Jon also remembers staying the night with his grandpa. A night at L.W.'s house nearly always meant Jon got to go out to eat for supper. Jon smiles while he reminisces about his boyhood days with L.W. The relationship between a grandfather and grandson is a special one. Lucky is the boy who gets to spend his days trotting to keep up with grandpa's long strides across the pasture, barnyard or cow lot. Even as L.W.'s body aged, he was a strong, barrel-chested man. One leg hurt him frequently, but other than that ailment and his chronic hard hearing, L.W. was a healthy, active grandfather.

Jon says, "One of the places we often went to eat on the weekends was the Railhouse restaurant in Centralia. We usually stopped to pick up his girlfriend at the time."

Jon explains that after L.W. became a widower he always had a girlfriend. However, it wasn't the romance he sought

L.W. holds his grandson, Jed Angell. Notice the large, clunky hearing aid L.W. has to wear.

out, but rather the company. L.W. simply didn't like being alone. He enjoyed going out for dinner, watching ball games and spending time with people. L.W. always made his grandson feel perfectly welcome on his dates and never a tag-along.

Jon says, "One time, his girlfriend wanted to go and see a movie pretty bad. It was showing at the Be Be Drive-In, so the three of us went out to see it. I remember the movie was long and kinda boring for me. But I'm pretty sure that was the first time I saw *Gone with the Wind*. Most of the time after we went to dinner or the movie, L.W. and me would drop off his date and head back to the quiet house. Well, at least it was quiet until we came in. Then, he played the TV really loud and sat in his chair, often smoking a cigar and reading his paper."

Because of L.W.'s poor hearing and the roaring volume on the television, L.W. couldn't hear his telephone. To remedy this situation, L.W. had an extra loud bell, similar to an old school bell, wired into his phone line. Sometimes even this wasn't enough to be heard over the TV. So he also had a light bulb contraption that sat on top of the TV. If the phone rang, the light would come on, too. After all the extra effort, L.W. was able to answer the phone, even though he was terribly hard of hearing.

Jon says, "When I would stay the night at L.W.'s house, it was my job to get up and fix breakfast. I always fried either sausage or bacon. L.W. liked his eggs cooked over easy. One of the first things Mom, Joan, taught me to cook was bacon and eggs. This came in very handy. Joan taught me to cook the bacon nice and crisp but at L.W.'s, I had to be retrained. He liked his bacon soggy and a bit greasy. I never knew if that was because he liked the flavor better or what. But now I suspect it was because he didn't like the crisp bacon bits getting stuck in his dentures. So I learned how to make soggy bacon for him. He also taught me how to fry up the eggs in a cast iron pan he stored in the oven below the stove using the grease from the bacon."

Jon and L.W. didn't have extravagant times together. They enjoyed simple everyday activities perfect for a grandfather and his young grandson, like burning boxes and cooking eggs.

Jon says, "L.W. also went to Kansas City a couple times a year to buy clothing at the western market from wholesalers to sell in the store. I never went, but a weekend finally came when several of my cousins and I could go along."

Jon traveled to the auction at the Kansas City Stockyards with L.W. on several previous trips, but he was looking very forward to the new trip to the clothing market.

"We loaded up in the car," Jon says. "There were six of us: me, L.W., my cousins Scott and Lori, Lori's friend Carla

(Buck) Armontrout and L.W.'s girlfriend. I was in seventh or eighth grade at the time."

The group drove to Kansas City. Jon recalls that L.W. slept most of the ride. When they arrived at the clothing market, they pulled up to the front door so everyone could get out. When someone woke L.W. up, he seemed disoriented and confused.

Jon says, "He sort of slurred his speech and said he needed to go to the bathroom. So, we helped him out of the car and he had trouble with his balance. Scott and me got on each side of him and we helped him into the bathroom. We knew something bad was happening and were talking among ourselves about what to do for L.W. While we were in the bathroom, Lori was at the payphone calling home to see what to do. On our way in, we passed the security guards and the ladies working the registration table. They could see he was in distress, but at the time they weren't of much help, it seemed. Several minutes passed and he wasn't getting any better."

L.W. heard them discussing the idea of taking him to the hospital and he didn't like it. He sternly slurred, "Take me home."

Jon was the youngest in the group and he tried to speak up and add his opinion to the chorus of panicked voices.

Jon says, "I repeated over and over again, 'We need to take him to a hospital!' I remember being very frustrated because no one was listening to me."

The eldest in the group made the tough decision to start driving back to Centralia. The day at the clothing market was obviously canceled.

Jon says, "They told me to get in the car. L.W.'s girlfriend drove to get us out of the city to head home. Later, she got upset and she put Lori to driving after we were out of the city. Lori called home again and finally got a hold of her father, Buddy, who told her to drive straight to Columbia and the Boone County Hospital."

Lori remembers driving on I-70, speeding toward

Columbia with four passengers across the back seat and L.W. sitting in the passenger seat.

From time to time, L.W. would look across the car at Lori and say, "Take me home. Take me home."

When they arrived at the hospital, the situation wasn't good. L.W. suffered a massive stroke. After that day, he was paralyzed and bedridden. His speech was limited, and at times he wouldn't speak at all.

Jon says, "The doctors told us that it didn't matter. The stroke had been so severe that going to the hospital in Kansas City wouldn't have helped him. But to this day, I

L.W. holds two grandsons, Tim Boender and Jed Angell. Tim is Rosemary's son and Luther's nephew. Jed Angell is Luther's son.

don't believe that. It was just a bad situation. I've always thought that it might have had a much different outcome if we had gone to the hospital in Kansas City two hours sooner. Would they have been able to release the pressure on his brain? Or maybe drain some of the blood away? I felt guilty I couldn't get him help right away. His commanding and authoritative personality swayed heavily the decision to take him home. That is just the way it was."

I talked to them separately but found it interesting that Buddy also expressed similar regrets. Neither Buddy nor Jon talked about their similar feelings until my prodding questions brought up the tough day.

Buddy said, "We made the wrong decision. We should have taken him to Kansas City right away."

It is impossible to know the outcome if L.W. had been taken on a short drive to a Kansas City hospital right after the stroke. Instead, we know that L.W. never got his wish of going home. From time to time, he might squeeze a visitor's hand and mumble a few words. The rest of L.W.'s life passed in the nursing home. Although sadly, it wasn't much of a life. They were simply six long, empty, sad years.

The stroke robbed L.W.'s body of conversation and movement, but he still had great physical strength. He lived for six years and passed away in June of 1987.

Until the time that I began writing our family stories down, this story was never mentioned – a significant fact in a family that loves to swap stories. I learned that this was a sad, hard time for all my relatives. As I asked questions, each member of the family still winced at the memory – brief words, watery eyes, revisited wounds.

Jon says, "It was a very discouraging time. I can't think of a worse way to go. I wouldn't wish that on anyone."

Justin echoes his brother's thoughts, "It was a very sad time. Going to visit him was really hard. It wasn't my grandpa laying there anymore."

I read an old newspaper article from the Columbia

Livestock Auction days. The reporter asked L.W. something about the future and he said planned to work at the livestock auction until he died. He did not get that wish either. In his tragic ending, L.W. left a final legacy to the Angell family. Everyone – cousins, aunts, uncles and grandparents – hopes for a death that mirrors Louise's rather than L.W.'s. She died peacefully in her own bed on a Sunday afternoon. He wasn't so fortunate.

Luther Washington Angell, Jr.
"L.W." or "Junior Angell"
June 26, 1910 – June 3, 1987

CHAPTER 47

ONE TOUGH DILEMMA

"Damn," Luther grunts as he reaches down to pull on his boots. By the time he reached his seventies, his belly had become so large that he couldn't reach the leather loops on the tops of this cowboy boots. As a younger man, he simply bent down and pulled them on. Now, he uses two silver boot hooks with red wooden handles. He inserts the hooks into the loops on either side of his calf and then leans back on his bench pulling the red handles mightily – errrrrumph – a boot squeezes up and around his heel. Five minutes later, Luther is winded, but both his boots are on!

"I'm going up to the sale," Luther yells to Joan on his way out of the house, letting the screen door slam shut behind him.

The boot hooks are just the latest nod to Luther's advancing age and dramatic change in body mass. One morning before a sale, Luther walked around the back of the livestock auction in Columbia inspecting the cattle. He peered into each pen, checking to see what was going to sell that afternoon. From time to time, someone yelled, "Heads up" to give Luther and the others some warning that a few more cattle were coming down the alley to be penned.

One group of particularly mean cattle started running down the alley. The employees yelled "Heads up" and "Get on the fence, they're wild!"

Luther looked up and he could see that he needed to climb up the fence in a hurry! A mean-looking bunch of yearlings were coming right for him. He began to climb the fence, but as he did, a terrible thing happened. His quick movement and matured body shape caused his pants to drop suddenly from his waist down to his ankles! With his pants at his feet, Luther was hobbled like a horse. He looked down the alley. The cattle were running closer every moment.

Luther says, "I didn't know what to do! Did I have time to reach down for my pants? Or should I just try to climb up the fence with my pants around my feet?"

It was a tough dilemma for Luther and he didn't have much time to decide! Disregarding all his dignity, he hopped, scratched and scrambled to make small steps up each six-inch wide board of the fence. The cattle passed, and the immediate danger was over.

Luther stood two or three bars up the fence with his underpants exposed for all to see. It was obvious that Luther preferred briefs over boxers. Everyone snickered and giggled as Luther left the back of the barn and headed toward the main sale arena. By the time he made it to the front of the barn, news of his traumatic event had already reached the auction block.

That weekend, Luther made a new purchase at the western store. It was something he'd been avoiding for years, but now Luther realized this item just might save his life. He decided the time had come. He broke down and bought his first pair of suspenders. In fact, he bought three pair – brown, black and blue.

Luther says, "I always wear them. I haven't missed a day since!"

CHAPTER 48

THE BIG COW WRECK

One Thursday morning in 2009, Jon sat at home doing book-work. He noticed a call coming in from his nephew, Jensyn, so he answered.

Jensyn said, "Uncle Jon, a cow has knocked Luther down. Can you come to the Grubbs' farm and help me get him up?"

Jon recalls that from the tone of Jensyn's voice, most people would have thought Luther's situation wasn't serious. However, Jon realized that Jensyn is one of the most calm, "even keel" people he had ever met.

Jon says, "Luther was seventy-five years old and Jensyn used the phrase 'help get him up'. Those were two signs that this situation was likely worse than Jensyn's calm tone implied."

For several years, it seemed to us that Luther might avoid the burdens of old age. He still wore his cowboy boots and worked outside daily. At seventy-five, his health was excellent; Luther was in prime condition. Of course, there were a few exceptions: he couldn't hear well, he had a big belly and issues with high blood pressure. With the exception of needing suspenders, Luther was aging in prime form.

Within five minutes of the phone call, Jon arrived at the Grubbs' farm and found Luther propped up in the mud, leaning against a fence behind the barn. Jensyn and Jon grasped Luther's belt and tugged him upwards. Once he was standing, they led him to a feed trough where he could sit down.

Jon went to fetch the truck. "Luther was unable to move his right arm and in a good deal of pain. It was obvious we had – at the least – a broken arm."

As Jensyn and Jon loaded Luther into the pickup, Jon asked what happened. Between Jensyn and Luther, Jon was able to get the full story.

Jon says, "It seems that Luther had enlisted his grandson to help sort through some cows to sell. The slaughter cow market had recently improved, so he decided it was a good time to market some cull cows. As they worked, a wild renegade emerged from the bunch. She lowered her head and targeted Luther – right in the seat meat! The mean ol' cow made contact with her target at a good speed. Then she raised her head and launched Luther skyward, flipping him over backwards to land on his head and shoulder. As he landed, the old hussy turned around and came at him again. She proceeded to maul the old guy in the mud while he was down! Jensyn was able to muster his best rodeo clown imitation to distract the brute, convincing her to rejoin the bunch."

Luther explained to Jon, "I had no idea she was coming at me. But, the moment that old cow hit me from behind, I knew just which one had gotten me!"

Jon says, "Now don't you think that says something about Luther's cowboying skills? He knew his cows good enough to identify one from the herd while being launched in the air and gazing at the sky!"

As Jon and Luther pulled out of the lot and headed toward the emergency room, Luther barked orders to Jensyn. He explained where to put the cows and how to fix the gates. Jon thought all of this was a good sign.

Jon says, "Then I had to listen to grumplin' and gasps at every bump to the hospital for thirty miles. I never noticed how many potholes were in that road until I was riding with the newly created bump-o-meter. As Luther complained about the drive, I reminded him, 'You missed the perfect excuse to hire an ambulance. If you weren't such a tight ass and hired an ambulance, by now you could have had some pain killers.'"

Jon and Luther arrived at the hospital and Luther seemed to be doing well. A second positive sign about his condition was when his trademark sense of humor seemed to be intact.

Luther told the E.R. nurse, "You know, if I had landed on my fat belly instead of my arm...I wouldn't be here!"

Jon remembers that the hospital staff knew just how to handle Luther's teasing and joking. After seeing the E.R. doctor, a nurse told Luther that they had called another doctor in to check him out. Luther asked the nurse who the second doctor would be.

Without missing a beat, the nurse replied, "He is our cow collision specialist."

Jon says, "Within a few hours, we got Luther all settled in. His regular doctor and family friend decided to keep Luther overnight. I think the doctor did this as a favor to my mother, Joan. Luther isn't known for being a problem-free patient. Once they moved him to a room, his need for humor emerged again, only this time it was exaggerated by a dose of pain-killing medications. He began contriving an elaborate ruse for his own amusement."

Joan was on her way to the hospital. When Luther found out she was on her way, he started in on his ploy.

He told the new attending nurse, "Now, it shouldn't be long before a cranky old woman will arrive here. She will be looking for me. Whatever you do, make sure you don't let her in my room. If you do, there will be big trouble!"

The nurse looked at her patient and replied, "Are you serious?"

Luther continued the stunt by saying, "Darn right I'm serious! My wife and I have had trouble. I'm not feeling up to it and I don't want to put up with her. I am warning you, if she comes in here, there will be big trouble!"

The nurse left Luther's room to set up a defense strategy. Within a few hours, Luther had everyone at the hospital – doctors, nurses and secretaries – right where he wanted them. The staff was prepared to turn away Luther's "cranky old woman". The only problem was that the only "big trouble" would be if his plan actually happened. Imagine if the hospital staff had turned away poor Joan?

Luther finally started feeling guilty. He let the nurse know that he had only been joking around. They couldn't believe that they were all prepared to send away his well-meaning wife!

When Joan arrived at the front desk on the appropriate floor, she told the secretary she was looking for Luther Angell's room. Before the secretary could type the name into the computer system, a co-worker offered, "Oh you don't need to look that one up – he is the funny old man in room 4222."

Just as Luther made an impression around the sale barns, he made quite an impression on the staff at the Boone County Hospital. Jon left the hospital confident that – although banged-up and bruised – Luther was going to be fine.

On Friday morning, Jon told a couple of the senior cattle buyers at the Eastern Missouri Commission Company, Bill Johns and David Thompson, about the big events of the previous day.

Bill and David made it clear that they were no longer interested in sorting cattle because, "There is no way that we can get out of the way anymore!"

They suggested that Luther follow their example.

Bill told Jon, "Let me tell you something. You hear all these folks talk about the golden years. Well, don't believe them. It is a big lie. If I had any golden years, they were back when I was eighteen to twenty-four!"

Bill had a lot of wisdom wrapped up in that statement. That week Luther found out that old, frail bones are more likely to break. A couple decades prior, Luther's big cow wreck wouldn't have been a big ordeal. However, at seventy-five, he found himself in his first cast with a broken arm. A cast for all of July and August was an unpleasant experience, but Jon reminded his father that "the golden years" are supposed to be about trying new things!

Jon says, "What is the definition of the golden years anyway? Aren't they supposed to be all about doing what you want to do and experiencing things you haven't gotten around to yet? In this case, I think my father Luther is wildly successful in his golden years!"

I asked Luther about the big cow wreck a few years later. By this time, he was almost eighty. He pondered for a moment and replied, "You know, I never really felt 'old' until that cow got me down. I don't think I ever really recovered from that one."

The ox cart.

CHAPTER 49

A LIFETIME COLLECTION OF PLAYBOY MAGAZINES

"When I die," Luther says to me, "my oldest son Jed gets the ox cart. Your great-great-great-grandpa made that for the 1904 World's Fair in St. Louis. It won best of show. He carved out the little oxen from wood and made the whole thing by hand. Since then, it always goes to the oldest child. Next, it will go to your cousin Jayci since she is the oldest."

"Well," I ask, "what about your collection of *Playboy* magazines?"

"What about them?" Luther says.

"I'm pretty sure that my father is hoping you'll give those to him. I think it is pretty rare for one person to have every single *Playboy* magazine ever published. Who knows, that collection might be more valuable than you think!"

At eighty years old, Luther smiles and finds it perfectly appropriate to continue his long-time subscription with the magazine. Over the years, it was actually my grandmother, Joan, who gave him the subscription for his birthdays and other holidays.

Luther points to his attic, "There's *Playboy* magazines up there from back in the days when the centerfolds were still wearing clothes. It wasn't very much clothes, but they had some on just the same. Now that I think of it, Marilyn Monroe might have been the very first centerfold that magazine ever did! You know something? The three prettiest women I ever saw were in those magazines."

I cringe and act annoyed.

I think to myself, "How does a granddaughter wind up talking to her eighty-year-old grandpa about his favorite *Playboy* centerfolds?"

Really though, I'm not annoyed. In fact, I don't know how I got so lucky. Some grandpas just want to give lectures and reminisce about the good ol' days gone by. Instead, I get to spend time talking to my wild, crazy grandpa about his funny stories and silly pranks. Luther interrupts my thoughts with his list of voluptuous women.

"The first woman was Patti McGuire," Luther says. "They had her posing next to a juke box."

Patti was featured in the November 1976 issue of the magazine. Luther's memory was correct. The first edition of *Playboy* came out in December of 1953, featuring a fully clothed – dare I say – classy photo of Marilyn Monroe on the cover. Things at *Playboy* had changed by the time Patti was featured. A much more scandalous photo is featured on that 1970s cover.

"The second most beautiful woman I ever saw was Demi Moore, back when she was young. I don't think she was a centerfold, but boy was she pretty! And the third most beautiful woman was Donica, Danica…what's her name? The race car driver."

For one moment Luther finally sounds like the senior citizen he is – forgetting someone's name.

I suggest, "Danica Patrick?"

"Yep!" Luther grins, "That's her! She wasn't a centerfold either, but she sure is cute! Those are the three most beautiful women I ever saw in that magazine. Every edition is all up there in my attic."

I roll my eyes again.

Luther's dog Kip shows off. Joan says, "Our family has loved two dogs over the years. The first was Kip and the second was Ginger."

L.W. "Pawee" Angell, Jr., and grandchildren.
Left to right: Lori (Angell) VanMaanen, Tim Boender, Justin Angell, Jed Angell and Todd Boender, circa 1965.

Luther "Papa" Angell and granddaughters sharing ice cream.
Left to right: Emily Angell, Papa, and Rebekah Angell, circa 2004.

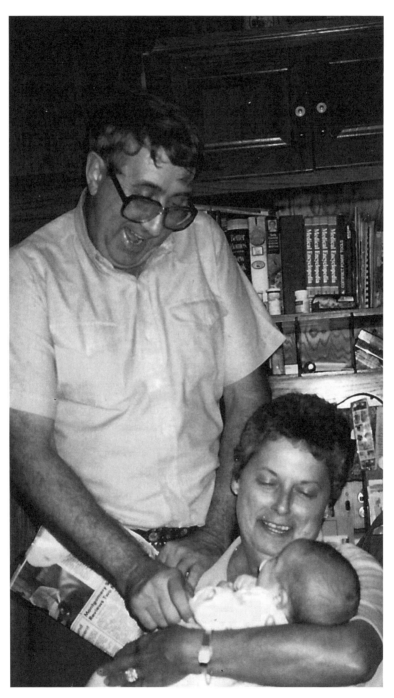

Luther tickles a new grandbaby.

Luther and Joan are surrounded by their children and grandchildren, one of the true joys of the so-called "Golden Years."

EPILOGUE

BIG FISH AND SECONDHAND LIONS

I stop by Luther and Joan's house unannounced, but I don't feel bad. They always welcome visitors, especially their grandchildren. Luther is down in the bedroom watching the movie *Secondhand Lions*. I sit on the end of the bed and watch it with him. This movie is about two old men who have a strong friendship. The old men are in their late sixties or early seventies. Most of society has written them off, saying their best days are past.

But these two put up a fight.

Their children and neighbors think that they should be living in a nursing home, but they disagree. One gets into a fight with several teenagers—and wins! They order a "used lion" and turn it out on their ranch. Once, they meet traveling salesmen on the porch with a loaded shotgun. I watch the end of the movie with Luther. The credits at the end of the film begin to play, and I look over and notice that he has tears streaming down his face.

He didn't hide his tears from me, but simply said in a blubbery voice, "I love that movie."

Out of respect, I pretend I don't see his tears, "I didn't know you like that movie. I like that one, too."

He says, "Yep, it is one of my favorites. I watch it every time it comes on TV."

Luther's unexpected tenderness towards the two characters doesn't surprise me. In his own way, Luther is just like those two men. He won't stop living his life just because he is "old."

One Sunday afternoon, my cousins, aunts, uncles, grandparents and I are all enjoying a family lunch. We grandkids have learned to visit and tell tales right alongside our parents and grandparents; we no longer look forward to watching TV.

After the meal, Luther announces, "I hope one of you kids finds me face down in the cow lot."

At first, I am shocked. I imagine him getting run over by some wild steers and badly hurt – a more dangerous and tragic version of his big cow wreck a few years ago.

He continues, "What I'm saying is, I hope I die with my boots on. That's all I want."

Normally, I'd be horrified to find someone dead in a muddy cow lot. But I start to hope this is how we find my grandfather. Like the men in *Secondhand Lions*, Luther does not want to spend his days waiting around to die. He will keep buying cattle, eating steak, drinking Canadian Club and watching his grandchildren play in ball games.

"And, I want my funeral to be a party!" he laughs. "Hire a band! Put them on the back of a flat bed trailer and they can lead the processional. Forget about going to the cemetery.

Just go on out to the country club and order everyone a round of drinks – that's when all the really good stories are going to come out."

"Ya!" Jon adds, "Rather than having a memorial line, we'll put Sierra in a booth with her recorder. We'll say, 'You can't come into the party until you tell a funny story about crazy ol' Luther!' That can be another book!"

We are all laughing and smiling. Then Luther throws out one more punch line to his crowd.

"Unless Joan goes first," he yells while winking at Joan. "If she dies, then I can tell all my good stories! You know she won't let me tell most of them."

Joan smiles, knowing she's kept some secrets safe in the past. Luther pats her on the leg sweetly.

"Hun," he says, "let's tell them the one about the green army blanket. What do ya think?"

"No!" she says. "Not until somebody finds me dead in a cow lot!"

"A man's body may grow old, but inside his spirit can still be as young and as restless as ever. And him – in his day, he had more spirit than twenty men."

– Garth, *Secondhand Lions*

ACKNOWLEDGMENTS

Of course, I am grateful to all of the family members who took their time to tell and retell these stories to me.

I am thankful to my father for being the first person to encourage me to record our family's history. His example showed me how to appreciate storytelling and he encouraged me to take the time to translate our family's oral history to a written history.

Thank you to my husband, John, who sometimes has to fix his own meals, start laundry *and* care for the cattle while I write.

Lucy, my dearest friend, thank you for all of the times in the past seven years when you asked, "How is the book coming?" Your voice was always full of faith; you knew I would get this book done before I did.

I'm grateful to Linda M. Hasselstrom, who encouraged me to write my own voice into this book. She offered comments and guidance on several, very long drafts and allowed me to write for several days at Homestead House.

Thank you to Dr. Laurel Wilson, who encouraged and mentored this project while I attended the University of Missouri. Her advice and introduction to "long-form interviews" was helpful. Financially, I'm also thankful to the CAFNR Undergraduate Research Grant; I spent much of the time for that research project completing these interviews.

Readers of *The Cattlemen's Advocate*, I appreciated your positive commentary while we published some of these stories in serial form. A special thanks to Chuck Herron and Jon Angell for carefully and tenderly editing those first stories for the newspaper. I always enjoy working with you both.

Thank you to Jon Angell for mentoring the project from infancy to completion, and to Jed Angell, the family genealogist, for your assistance with important dates, names and helping me understand our family tree.

And finally, thank you to the South Dakota winters of 2013 and 2014. You were brutal, keeping me indoors and isolated *and* you gave me months to work on this book. If I lived in Hawaii, this project might have stayed a dream.

FURTHER READING

The following books were helpful to me as I wrote this book:

Choate, Wade. *Swappin' cattle*. (4 ed.). San Angelo: Newsfoto Publishing Co, 1990.

Collier, Gaydell. *Just beyond harmony*. Glendo: High Plains Press, 2011.

Blew, Mary Clearman. *Writing her own life: Imogene Welch, western rural schoolteacher*. Norman: University of Oklahoma Press, 2004.

Walls, Jeannette. *Half broke horses: A true-life novel*. New York: Scribner, 2009.

ABOUT THE AUTHOR

Sierra Shea is a writer and ranch wife from De Smet, South Dakota. She is the author of "So God Made a Farm Wife" and "So God Made a Ranch Wife". A Missouri native, Sierra began writing as a columnist for *The Cattleman's Advocate* as a sophomore in high school. Today, she  writes and blogs for a variety of clients. She is passionate about preserving the heritage of rural people by writing their stories. Sierra blogs about her writing, ranching, and quilting at www.sierrashea.com.